THE COMPLETE ILLUSTRATED GUIDE TO

ROCKS
OF THE WORLD

THE COMPLETE ILLUSTRATED GUIDE TO
ROCKS
OF THE WORLD

A practical directory to over **150** igneous, sedimentary and metamorphic rocks

John Farndon

CONSULTANT: DR ALEC LIVINGSTONE,
THE NATIONAL MUSEUMS OF SCOTLAND, EDINBURGH

LORENZ BOOKS

This edition is published by Lorenz Books
an imprint of Anness Publishing Ltd
Blaby Road, Wigston, Leicestershire LE18 4SE;
info@anness.com

www.lorenzbooks.com; www.annesspublishing.com

If you like the images in this book and would like to investigate
using them for publishing, promotions or advertising,
please visit our website www.practicalpictures.com
for more information.

Publisher: Joanna Lorenz
Managing Director: Helen Sudell
Project Editor: Catherine Stuart
Production Manager: Steve Lang
Consultants: Dr Alec Livingstone and Dr John Schumacher
(Department of Earth Sciences, University of Bristol, England)
Book Design: Nigel Partridge and Michael Morey
Cover Design: Nigel Patridge

© Anness Publishing Ltd 2012

PUBLISHER'S NOTE

The author and publishers have made every effort to ensure
that all instructions contained within this book are accurate
and safe. Persons handling and collecting rocks and mineral
specimens do so at their own risk. This book offers guidance
on the basic safety precautions to take when handling,
identifying or collecting rock and mineral specimens; however,
we cannot accept any responsibility for any injury, loss or
damage to persons and property that may arise as a result of
these activities. It is especially important that collectors observe
the protective laws applied to a particular geological site before
entering the site or removing rock and mineral specimens, and
note that these may vary from region to region, and from
country to country. Useful information can, and should, be
obtained from local government and environmental agencies
before visiting any geological site.

PICTURE ACKNOWLEDGEMENTS

Note: T = top; B = bottom; M = middle; L = left; R= right

The publishers would like to thank Richard Tayler, who provided
the majority of specimens photographed for this book from his
own collection, with additional material from his frjends and
associates.
The specimens were photographed by Martyn Milner. In
addition, the Department of Earth Sciences, University of
Bristol, supplied the specimen photographs of greensand (91B),
doloritic limestone (101T) and fulgurite (121B). The School of
Earth, Atmospheric and Environmental Sciences, University of
Manchester, supplied the specimen photographs of boninite
(57T), lapilli (66T), tephra (66B), breadcrust bomb (67T), melilitite
(69B), monzonite (73B), monzodiorite (78B), charnockite (107B),
greenschist (108T).

Peter Bull drew the geological artworks.

Geological photographs were provided by the following:
2 (Granite rock exposed at Tarn (Midi-Pyrénées), France),
Nature Picture Library; 7TL, Natural History Museum.
Understanding How Rocks Are Made: 19BR, 27TR & 40–1,
British Geological Survey; 44BL, David Huston, Geoscience
Australia; 17TR, Marli Miller; 23TR, Natural History Museum;
32BL, Nature Picture Library; 13BR, NHPA; 10BL, 18, 22, 25TR,
27B, 29T, 30BL, 32TL, 36TR, 36BL, 38T, 40, Oxford Scientific;
13TL, Science Photo Library; 34TR & 44T, Still Pictures; 27TL
& 28M, Department of Earth Sciences, University of Bristol;
45B, Andrew Wygralak, Northern Territory Geological Survey.
World Directory of Rocks: 107, 113 & 119, British Geological
Survey; 79, Peter Frank © 2005 Canadian Museum of Nature,
Ottawa, Canada; 103, Earth Observatory, NASA; 61, Nature
Picture Library; 67 & 97, Natural History Museum; 111 & 115,
Oxford Scientific; 93, Anthony G Taranto Jr, Palisades Interstate
Park; 57 & 121, Science Photo Library; 59, Still Pictures; 81,
David L Reid, Department of Geological Science, University of
Cape Town, South Africa; 55 & 117, Ian Coulson, University of
Regina, Canada.

CONTENTS

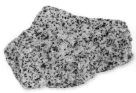

INTRODUCTION

Back in the early 19th century, when the whole idea of building a collection of rocks was just getting under way, the famous Scottish poet Sir Walter Scott (1771–1832) described geologists thus, "Some rin uphill and down dale, knappin' the chucky stones to pieces like sa' many roadmakers run daft. They say it is to see how the warld was made!" Even today many people consider hunting rocks and minerals a rather strange activity.

Yet go to a beach or any stream where the water has exposed sand and gravel banks. You will see stone upon stone lying there. To start with, they probably all look dull and grey. But look closer and you begin to see subtle differences in colour. One might be pale cream. Another mottled brown. A third slightly stripy. To the untrained eye, they're still just stones, but to an experienced rock hunter, each has its own fascinating story to tell, manifest in these subtle markings.

Below: Pegmatites are the apple of every rock collector's eye. They are formed at the fringes of igneous intrusions from mineral-rich melts. The gigantic crystals they embody have a high content of water, which means that they can grow very big, very fast. It is not always easy to be precise about crystal content, but many pegmatites contain quartz, feldspar, mica, tourmaline, and in some cases rare gems. One of the largest crystals ever produced by a pegmatite was 13m/42ft long!

The pale cream stone could be limestone. Look at it through a magnifying glass, and you begin to see it is made from tiny grains with shiny surfaces, each a crystal of calcite precipitated out of tropical oceans hundreds of millions of years ago. Here and there in the stone you may actually see fossils of the sea creatures that swam in these ancient oceans.

Your rock hunter might go on to tell you that the mottled brown stone is granite – then show how through the magnifying glass you can see three different minerals. There are tiny black flecks of mica, glassy grains of quartz, and yellow feldspar – all forged in the fiery heat of the earth's deep interior millions of years ago. The stripy stone could be a schist, a rock formed when other rocks came under such intense pressure from earth movements that the crystals in them broke down and were made anew in different forms – squeezed into stripes by the pressure. A magnifying glass might reveal tiny red spots embedded in the rock. A rock hunter might identify these as garnets or even rubies, tiny versions of the beautiful gems that kings have fought and died for.

With all this and much more in just three stones picked up on a beach, it is not surprising that many people have become hooked on rock collecting. Some specimens are worth hanging on to for their sheer beauty: if you're lucky enough to find a pegmatite, you might find it flecked with pink beryl, blue amazonite feldspar, or smoky quartz. Others are fascinating for their shape and texture. Nature's own marbles,

Right: Although many types of limestone are rich in the deposits of shell creatures, few entomb fossilized remains as visually as this bryozoan limestone specimen. Bryozoans are ancient moss-like creatures that sometimes form colonies of tiny tubes.

Above: Some specimens only reveal their true beauty when ground and polished like these specimens. At the top is the rock orbicular, a special variety of granite. Below are isolated nodules of obsidian, 'Apache tears'.

smooth obsidian 'Apache tears', are created from the rock perlite in a series of chemical reactions; porous rocks such as dripstones appear as a 'frozen' embodiment of the calcite-rich water that created them. In addition, rocks add immense value to industry – some are sources of the ores that provide us with metals, or of building materials. Yet even when neither beautiful nor valuable, stones have a fascination because of what they tell us about the geological processes that continue to shape the Earth.

Above: The erosion-resistant sandstones that form the massive cliffs at Mesa Valley National Park, Colorado, were used in the construction of the famous Cliff Palace, a dwelling built by the Ancient Pueblans of south-western USA.

Rocks through the ages

Rocks have long played a part in human history. Long ago, our ancestors were chipping the edges off hand-sized pebbles, perhaps to use as weapons. At least two million years ago, hominids began to use flints to make two-sided hand-axes, which is why the first age of humanity is known as the Stone Age. Finding good flints required a considerable practical knowledge of

Below: Each of these rocks has a story to tell. Top left is a dripstone formed in caverns from minerals dissolved in rainwater. Top right is siltstone, a gritty-textured rock often interlayered with darker mudrock, and rippled like the river currents that deposited the sediments. At the bottom is a limestone imprinted by ancient corals.

geology. Few people today would have a clue where to look for flints – yet these Stone Age people knew, and even dug mines to get at them underground.

Other rocks have acted as the building blocks of the modern age. Widely available sedimentary rocks such as sandstones and limestone have contributed to the construction of entire cities, from magnificent palaces to towering office blocks. Despite its name, sandstone is a very tough rock, not easily eroded – the world renowned monolith Uluru in Northern Territory, Australia, is composed of arkose sandstone. Feldspar-rich sandstone, or brownstone, can be seen in the facades of Boston and New York

apartments, and this group of durable rocks continues to be mined as valuable building and paving material across the globe. The natural, glassy-smooth texture of marble, metamorphosed carbonate rock, is another wonder still. The compaction of calcite grains, which occurs deep beneath the mountains under immense pressure, produces a rock with a glassy yet soft texture and translucent glow, which has been prized by sculptors since ancient times.

The rock hunters

It was not until the late 18th century that geology emerged as a science, pioneered by the great Scottish geologist James Hutton (1726–97). In Hutton's day, most people still believed the Earth was just a few thousand years old. Hutton realized it is much, much older – it is now thought to be 4.5 billion years old – and that the slow processes we see acting on the landscape today were quite enough to shape it without invoking great catastrophes, as others did. Hutton demonstrated that

Above and left: Even rather colourless rocks may contain valuable gems. This kimberlite specimen (top) is studded with a diamond; the solid red mass in this schist (left) is the red variety of corundum known to the world as ruby.

landscapes are worn away by rivers and that sediment washed into the sea forms new sedimentary rocks. He also saw how the Earth's heat could transform rocks, lifting and twisting them to create new mountains. So the world is shaped by countless cycles of erosion, sedimentation and uplift – each new beginning often clearly marked by breaks in the rock sequence called unconformities.

Inspired by Hutton's ideas, more and more geologists ventured out into the field to explore rocks and the story they tell. Geology became a popular pastime for many Victorian gentlemen. A huge proportion of the rock and mineral species we know today were first identified and named by Victorian specimen hunters.. They included Charles Darwin, who brought his knowledge of geological history to bear in formulating his ground-breaking theory of evolution.

As the following pages show, our knowledge of geological processes has developed tremendously since those early days, and professional geologists are aided by a range of sophisticated equipment. Yet the amateur armed with a few basic tools and a sharp eye can find fantastic specimens. It is the aim of this book to help in this search.

UNDERSTANDING HOW ROCKS ARE MADE

Rocks are the raw materials of the landscape: every valley, hill and mountain peak is made of rock. Yet each one is also a clue to the Earth's history, for their characteristics depend on how and where it formed – whether forged in the heat of Earth's interior, transformed by volcanic activity and the crush of moving continents, or laid down in gently settling layers on the sea bed.

Climb any mountain away from the city and gaze out on the landscape, and you see rivers winding down to the sea, hills and valleys, forests and fields. It seems a timeless landscape, but in terms of the Earth's history it is very young. The fields may be no more than a few centuries old, the forests a few thousand years old, and even the hills and valleys no more than tens of thousands of years old. Yet the history of the Earth's surface and the rocks which make it dates back over four and a half billion years. In geological terms, then, the landscape we see today is just a fleeting moment.

Geologists first began to understand the ever-changing nature of the Earth's surface about 200 years ago, as they started to realize just how old it was, and how it was shaped and reshaped continuously by the power of such forces as water, earthquakes and volcanoes, which have worn mountains down and risen new ones up. Yet it is only in the last half century that they have learned just how dynamic the Earth's geology really is, with the discovery that the entire surface of the planet is on the move – broken into 20 or so giant, ever-shifting slabs called tectonic plates. The movement of these plates is slow in human terms – barely faster than a fingernail growing – yet over the vastness of geological time, it is momentous: able to shift continents and oceans right around the globe. Tectonic plate theory has revolutionized geologists' understanding of how rocks are made and remade over again, how mountains are built and knocked down, why volcanoes erupt, why earthquakes happen and much more besides.

Left: The toughest rocks resist erosion by weather, water and time when everything else around them has worn away. This isolated peak forms part of the iconic geology at Monument Valley, Colorado Basin, USA. The scenery is distinctive for its 'buttes' – towering sedimentary rocks composed of sandstone and shale, the tougher outer rock protecting softer sediment underneath. As the softer rock is exposed, it crumbles away, causing the vast expanses of space that typify this ghostly landscape. The reds and blues that stain the Valley's rocks come from deposits of iron oxide, or manganese oxide, respectively.

INSIDE THE EARTH

The Earth under your feet may seem solid, but recent research has shown that its interior is far more dynamic and complex than anyone ever thought. Beneath the thin rocky shell called the crust, the Earth is churning and bubbling like thick soup.

Half a century ago, scientists' picture of the Earth's interior was simple. In some ways, they thought, it seems like an egg. The outside is just a thin shell of rock called the crust. Immediately beneath, no more than a few dozen kilometres down, is the deep 'mantle' were the rock is hot and soft. Then beneath that, some 2,900km/1,800 miles down is the yolk or 'core' of metal, mainly iron and nickel. The outer core is so ferociously hot that it is always molten, reaching temperatures as high as the Sun's surface. The inner core, at the centre of the Earth, is solid because pressures there are gigantic.

The key to this structure is density. The theory is that when the Earth was young, it was hot and semi-molten. Dense elements such as iron sank towards the centre to form the core. Lighter elements such as oxygen and silicon drifted to the surface like scum on water and eventually chilled enough to harden into a crust.

Above: Meteorites big enough to cause a crater like this one, the famous Meteor Crater in Arizona, strike the Earth so hard they vaporize on impact. But some small meteorites survive and provide vital clues to the composition of the Earth's interior.

Below: Blacksmiths know that iron only melts at very high temperatures, and scientists have deduced that the temperatures of the molten iron in the Earth's outer core, where it is under intense pressure, may rise to 4,500K/7,600°F. The solid core could even reach 7,500K/13,000°F!

Some heavy elements such as uranium ended up in the crust despite their density because they unite readily with oxygen to make oxides and with oxygen and silicon to make silicates. Substances like these are called 'lithophiles' and include potassium.

Blobs of 'chalcophile' substances – substances like zinc and lead that join readily with sulphur to form sulphides – spread up and out to add to the mantle. Dense globs of 'siderophile' substances – substances such as nickel and gold that combine readily with iron – sank towards the core.

The only real complications to this picture seemed to be on the surface, where the crust is divided into continental and oceanic portions. The crust in continents can be very ancient – some rocks are nearly four billion years old – and quite thick. Although

it is just 20km/12 miles beneath California's Central Valley, it is 90km/54 miles thick under the Himalayas. The oceanic crust, on the other hand, is entirely made of young rocks – none older than 200 million years and some brand new – and is rarely more than 10km/6 miles thick.

Listening in
Discoveries in the last few decades have forced scientists to re-evaluate this fairly simple picture. The problem has been to see inside the Earth. A Japanese ship launched in 2005 is now beginning to drill the deepest hole ever,

through the oceanic crust, hoping to reach the mantle, but this is barely scratching the surface. Yet there are other ways of telling. Astronomical calculations based on gravity tell us Earth's mass and show that the interior must be denser than the crust. Meteorites tell us a little about the mineral make-up of the interior, with the two kinds of meteorite, stony and iron, reflecting Earth's stony mantle and iron core (see Space Rocks, World Directory of Rocks). Similarly, volcanoes throw up materials like olivine and eclogite from deep in the mantle. Yet the main clues come from earthquake (seismic) waves.

Long after an earthquake, its reverberations shudder through the Earth. Sensitive seismographs can pick them up on the far side of the world. Just as you can hear the difference between wood and metal when tapped with a spoon, so scientists can 'hear' from seismic waves what the Earth's interior is made of. Seismic waves are refracted (bent) as they pass through different materials. They also travel at varying speeds, shimmering faster through the cold hard rocks of the crust, for instance, than the warmer, soft rocks of the mantle.

Density and speed

One thing seismology has revealed is that there is another way of looking at the crust and upper mantle. Although they may be chemically different, their 'rheology' is not – that is, they distort and flow in much the same way. Fast seismic waves show the top 100km/60miles of the mantle is as stiff as the crust, and together upper mantle and crust form a rigid layer called the lithosphere. Below the lithosphere, slower waves show that heat softens the mantle to form a layer called the asthenosphere. Tectonic plates are huge chunks of lithosphere that float on the asthenosphere like ice floes on a pond.

About 220km/140 miles down pressure stiffens the mantle again to form the mesosphere. Farther down, pressure forces minerals with the same chemical composition through a phase change (like ice melting) into a denser structure. So below 420km/260 miles olivine and pyroxene are replaced by spinel and garnet. Deeper down still, beyond 670km/420 miles, even higher pressure changes mineral structure again or maybe the composition, this time to give perovskite minerals, from which the bulk of the mantle is made.

The core boundary

Down through the mantle, seismic waves move ever faster. Yet at the Gutenberg discontinuity 2,900km/1,800 miles down, there is a drop in speed, marking the transition to the core along the Core-Mantle Boundary (CMB). The change is dramatic. In just a few hundred kilometres temperatures soar 1,500°C/2,732°F, and the contrast in density between mantle and core is even more marked than that between air and rock.

The transition zone in the mantle down to the CMB is called the D" (pronounced D double prime) layer, and has attracted a lot of attention. The outer surface of this layer is marked by valleys and ridges, and lab tests have shown that it may be made of a unique form of perovskite dubbed post-perovskite. In 2005, scientists detected an increase in speed below the D", suggesting that the outer rim of the core may actually be solid.

Research into this whole CMB zone may have key implications for our understanding of how continents move and volcanoes erupt, because these could be tied in with deep circulation of material in the mantle.

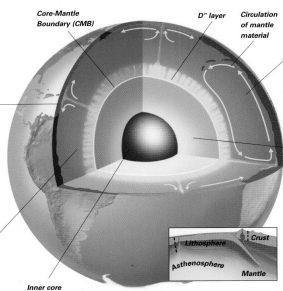

The Earth's interior
This globe reveals scientists' idea of the Earth's interior layers (not to scale), including the crust, mantle and core.

The crust *0–40 km/0–25 miles down is Earth's thin topmost layer, made mostly of silicate-rich rocks like basalt and granite. It is thinnest under the oceans and thickest under the continents. It is attached to the mantle's rigid upper layers, which floats in slabs on the soft mantle below.*

The lower mantle *is 670–2,900km/420–1,800 miles down. Here, huge pressures turn the lighter silicate minerals of the upper mantle into dense perovskite and pyroxene. Perovskite is the most abundant mineral in the mantle, and so in the Earth, since the mantle makes up four-fifths of the Earth's volume.*

Core-Mantle Boundary (CMB)

D" layer

Circulation of mantle material

The upper mantle *16–670km/10–420 miles down is so warm it is soft enough to flow. In the asthenosphere layer, below the lithosphere, pockets often melt to form the magma that bubbles up through the crust to erupt in volcanoes. The upper mantle is made mainly of the dense rock called peridotite.*

The Earth's core *2,900–6,370km/1,800–3,960 miles down is a dense ball of iron and nickel. The outer core is so hot, reaching temperatures of over 4,500K/7,600°F, that the metal is molten. The inner core is even hotter, up to 7,500K/13,000°F, but the pressures here are so great that the iron simply cannot melt.*

Lithosphere *(left) is the rigid outer layer of the Earth, broken into the tectonic plates that make up the surface. It consists of the crust and the stiff, cool upper portion of the mantle.*

Lithosphere

Asthenosphere

Crust

Mantle

Inner core

CONTINENTS AND PLATES

The division of the world into land and sea seems so natural and so timeless that it is difficult to imagine how it could be any other way. Yet the very existence of continents is remarkable and Earth is the only planet we know to have them. Their foundations are even more astounding.

Like Earth, Venus and many other planets and moons in the solar system have rocky crusts. But the crusts on these other worlds are made almost entirely of basalt, and are very stable and ancient, barely changed since they formed billions of years ago. Earth has a basalt crust, too. The crust under the oceans is basalt, for instance. But Earth's crust is mostly neither stable nor ancient. In fact, nowhere is the ocean crust older than 200 million years, and much is forming even now.

Even more unusually, the Earth has giant chunks of crust made from granite-like rocks such as granodiorite. It is these chunks of granitic crust that form our continents, light enough to ride high above the surface and form landmasses on partially molten matter.

Making continents
No one knows for sure exactly how or why these chunks formed, but their formation is clearly a very slow

process. The oldest pieces are almost four billion years old (the Earth is about 4.6 billion years old). Yet even now, granite crust still forms barely a quarter of the world's surface.

Basalt crust forms when magma (molten rock) from Earth's warm mantle cools and solidifies at the surface. Granite cannot form directly from mantle melts like this – only when basalt remelts, changing its chemistry and mixing in other substances met at the surface.

Geologists think this happens in two ways. The first is when hot, molten basalt magma wells up under the crust from the mantle, melting the crust to form granite magma. Less dense than basalt crust, the granite magma rises to the top before solidifying. The second process is when movement of the Earth's crust draws basalt crust back towards the mantle, remelting it and forming a new granite magma, which rises to form new continental crust.

This process initiated the evolution of the great continents that would support so much of Earth's living matter.

So just how does the Earth's crust move? The answer lies in the most powerful geological concept of the last century: plate tectonics. This is the idea that the Earth's rigid surface – all the lithosphere including the crust – is broken into 20 or so giant pieces of tectonic plates. These plates move slowly around the planet, carrying the continents and oceans with them.

Continental drift
The seeds for plate tectonic theory were planted in the early 20th century by a German meteorologist called Alfred Wegener. He had noticed the extraordinary way the west coast of Africa mirrored the east coast of South America. He also noticed amazing matches between widely separated continents of things such as geological strata and ancient animal and plant

The moving continents
Over the last 500 million years, Earth's continents have merged together then split and drifted apart. About 225 mya (million years ago) at the end of the Permian, all land was joined in one supercontinent geologists named Pangaea. Pangaea began to rift apart about the time dinosaurs appeared on Earth, so different kinds of dinosaur evolved in newly separated parts of the world. The maps here show the current continental shapes to help identify them, but in fact their shapes varied as low areas were flooded and mountain ranges rose up.

Permian 225 mya *During the Permian period all the world's landmasses moved together to form the giant continent Pangaea.*

Late Triassic 205 mya *During the Triassic, a wedge of ocean called the Tethys Seaway grew wider, elbowing into the east of Pangaea.*

Jurassic 150 mya *In the Jurassic, Pangaea began to split in various places, including the Tethys Seaway. South-western North America was flooded by the Sundance Sea.*

Cretaceous 80 mya *By the Cretaceous even southern Pangaea, called Gondwana, had split up to form today's southern continents. India began drifting northwards towards Asia.*

Present day *Over the last 50 million years, the North Atlantic has opened up to separate Europe and America, and India has crashed into Asia, throwing up the Himalayas.*

Above: In recent years, dramatic evidence of tectonic activity has been found deep under the sea. Fissures like this one in the East Pacific seabed, some 2,600m (8,500ft) below the surface, indicate the spreading of the oceans. Volcanic activity along these fissures warms the water and produces a rich mixture of chemicals, which creates a unique habitat for these white crabs, as well as other marine life.

fossils – especially from the Permian Period some 230 million years ago. Signs of ancient tropical species as far north as the Arctic circle only added to the impression.

Wegener guessed this was not mere coincidence. He realized that these matches might occur because today's separate continents were once actually joined together. He suggested that in Permian times they all formed one giant supercontinent which he called Pangaea, surrounded by a single giant ocean, which he called Panthalassa. At some time, perhaps some 200 million years ago Wegener thought, Pangaea split into several fragments and these have since drifted apart to form our present continental lands.

To many geologists of Wegener's time, the idea that continents drift around the world was ridiculous. The crust seems far too solid for this to happen. But over the next half century, the weight of evidence piled up. A key element was the discovery of grains of

Right: Looking at this bleak tundra on the Arctic island of Spitsbergen, it is hard to believe that lush tropical vegetation ever grew here. Yet fossils show that it did – not because the whole world had a tropical climate, but because Spitsbergen was once in the tropics, before continental drift took it way out into the chilly polar regions.

magnetite – a magnetic mineral – in ancient rock. These grains behaved like tiny compasses, lining up with the North Pole at the time when the rock formed. To geologists' surprise, these grains do not all point in the same direction. At first, geologists thought this must be because the magnetic North Pole had moved over time. Then they realized it was not the Pole that had moved, but the continents in which these grains were embedded, twisting this way and that. Geologists realized that with the aid of these ancient compasses or 'palaeomagnets' they could trace the entire path of a continent's movement through time – and this is how the maps below left were worked out.

Spreading oceans

A second key element was the realization that it was not just the continents moving but the entire surface of the Earth, including the oceans. Indeed, the continents are just passengers aboard the great, slowly sliding tectonic plates that make up Earth's surface. The breakthrough came in 1960 when American geologist Harry Hess suggested that the ocean floors are not permanent. Instead, he suggested, they are spreading rapidly out from a ridge down the middle of

the sea bed, as hot material wells up through a central rift, pushing the halves of the ocean bed apart. This does not make the crust bigger, because as quickly as new crust is created along the ridge, old crust is dragged down into the mantle and destroyed along deep trenches at the ocean's edge, in a process known by geologists as subduction.

Many were initially sceptical of Hess's idea, but evidence soon came when bands of magnetite were found in rocks on the sea floor in exactly matching patterns either side of the mid-ocean ridge. These bore witness to the spread of the ocean floor as truly as rings in a tree. Before long, the ideas of sea-floor spreading and continental drift were combined in the all-encompassing theory of plate tectonics.

It soon became clear that plate tectonics explained much more than coincidences of rocks and fossils. Earthquakes happen where plates shudder past each other. Volcanoes erupt where plates split asunder or dive into the mantle. Mountain ranges are thrown up where plates collide and crumple the edges of continents. In fact, the full implications of this revolutionary theory are only just beginning to be explored.

THE MOVING EARTH

Like a broken eggshell, Earth's surface is cracked into giant slabs of rock called tectonic plates. There are seven huge plates and a dozen or so smaller ones. The plates are not fixed, but ever shifting, breaking up and being made anew – almost imperceptibly slow in human terms, but dramatic geologically.

Tectonic plates are fragments of the lithosphere – the cool, rigid outer layer of the Earth, topped by the crust. The scale of some of these plates is staggering. They may be no more than 100km/60 miles thick, but some plates encompass entire oceans or continents.

The biggest is the Pacific plate, which underlies most of the Pacific Ocean. Interestingly, this is the only major plate that is entirely oceanic. All the others – in size order, the African, Eurasian, Indo-Australian, North American, Antarctic and South American – carry continents, like cargo on a raft. The remaining dozen or so plates are much smaller. But even the smallest of them, like North America's Juan de Fuca, is bigger than a country such as Spain.

It seems almost unimaginable that slabs of rock as gigantic as these could move. Yet they are moving, all the

The world's major plates
On this map showing the world's major plates, red arrows indicate the direction of movement.

time. Typically, they are shifting at about the pace of a fingernail growing – about a 1cm/0.4in a year, although the Nazca plate in the Pacific is moving 20 times as fast. Yet even a fingernail's pace has been fast enough to carry Europe and North America apart and create the entire North Atlantic in just 40 million years, which is a short time in geological terms. Accurate laser measurements can even detect the movements in a single month.

Why plates move
Scientists are not yet certain what it is that drives the plates, but most theories focus on the idea of convection in the Earth's mantle.

Inside the mantle, material is continuously churning around, driven up by the ferocious heat of the core, then cooling and sinking back. It was once thought that all these 'convection' currents moved in vast cells as big as the plates, and the plates simply rode on top of them like packages on

conveyor belts. Now more scientists are focusing on hot currents called mantle plumes that bubble up from the deep mantle, and in places lift the crust like a pie crust in an oven.

This may tie in better with other theories that suggest the driving force is the weight of the plates alone. Mid-ocean ridges are 2–3km/1–2 miles higher than ocean rims, so plates could be sliding downhill away from them.

Another idea is that plates are like a cloth sliding off a table. Hot new rock formed at the mid-ocean ridge cools as it slides away to the ocean rim. As it cools, it gets denser and heavier and sinks into the mantle again. It then pulls the rest of the plate down with it, just as the weight of a table cloth hanging over the edge can be enough to pull the rest off. It may be that all these mechanisms play a part.

Plate boundaries
Working out where one plate ends and another begins is far from easy. Seismic

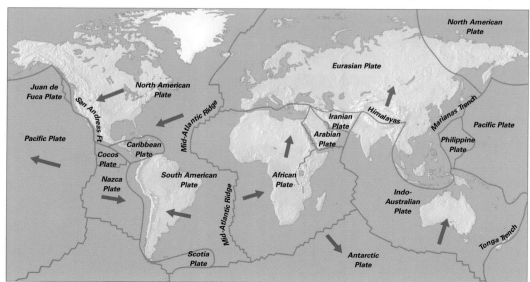

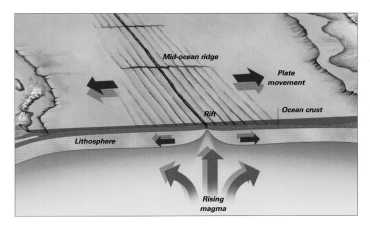

Divergent boundaries *typically occur in mid-ocean. Right down the middle of the Atlantic Ocean, for instance, there is a ridge about 2km/1.2 miles down on the sea bed, forming a jagged line where plates meet. In the middle of this ridge is a trough some 500m/1,640ft deep and no more than 10km/6 miles wide. This trough is where the plates are moving apart, spreading the ocean wider and wider. Beneath this trough, magma is welling up from the asthenosphere. Some solidifies on the underside of the crust, creating rocks like gabbro. Some oozes up into vertical cracks created by the pressure of the magma to form wall-like sheets of basalt rock called dykes. Some magma spills out on the sea bed and freezes in the cold water into blobs called pillow lavas. As soon as it forms, new oceanic crust moves away from the ridge, more magma wells up and new crust forms.*

profiling using earthquake waves helps build a picture but it is often vague.

One of the best indicators of a plate boundary is where earthquakes start. Plates generate earthquakes as they judder past each other so nearly all major quakes occur in belts that follow plate boundaries. In May 2005, a Japanese geologist discovered a previously unknown plate under Japan's Kanto just by tracing the origin of over 150,000 small earthquakes.

Other features that identify a plate boundary include long chains of volcanoes, ranges of fold mountains, curving lines of islands and deep ocean trenches. Some plates meet along the coasts of continents, like the west coast of South America. Coasts like these are called active margins. Other plates meet in mid-ocean. The continental crust is actually attached to the oceanic crust, and so its margin is passive.

Moving boundaries

All plates are on the move, some slow, some fast, and geologists identify three kinds of boundary, each with its own features. Divergent boundaries (see above) are where plates are pulling apart. Convergent boundaries (see below) are where they push together. Transforms are where they slide sideways past each other. Most transforms are short, linking segments of mid-ocean ridge. But some, like the Alpine Fault in New Zealand, link ocean trenches. The famous San Andreas fault in California lies along the boundary where the Pacific Plate is rotating slowly past the North American plate.

Plate boundaries can be quite short-lived geologically. Some ancient continental plates, for instance, were thrust together so hard they welded into one solid piece. There was once a plate boundary in Asia, north of Tibet, for example. Geologists detected evidence of this via a seismological image of the edge of one of the plates involved thrust far down into the mantle beneath the boundary.

Convergent boundaries *typically occur along the edges of oceans. Here, as plates crunch together, the lighter continental plate rides up over the denser ocean plate, forcing it down into the mantle in a process called subduction. Deep ocean trenches open up where the descending plate plunges into the asthenosphere. As it goes down, the plate starts to melt, releasing hot, volatile materials, water and even molten rock, creating magma. Plumes of magma rise through the often shattered, faulted edge of the overlying plate, and may erupt to create an arc of volcanoes along the edge, or maybe even volcanic islands. The over-riding plate often acts like a giant mudscraper, scraping material off the sea floor on the subducted plate and piling it high in a wedge called an accretionary prism. As the subducted plate shudders down, the vibration can set off earthquakes, creating what is called a Wadati-Benioff zone, after seismologists Kiyoo Wadati and Hugo Benioff.*

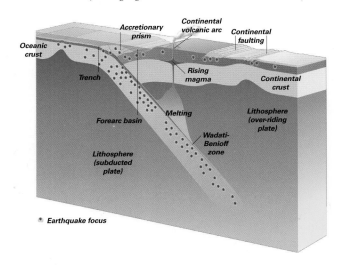

MOUNTAIN BUILDING

There are no more dramatic demonstrations of the dynamic power of the Earth's surface than mountains.
Lifting the vast bulk of rock tens of thousands of metres upwards into towering ranges such as the
Himalayas involves huge forces. Yet such forces have been deployed repeatedly in the Earth's history.

The remarkable thing about the world's biggest mountain ranges is just how young they are in geological terms. The Himalayas, the Andes, the Rockies and the Alps have all appeared within the last 50 million years. In other words, the dinosaurs were long dead before these so timeless-seeming mountain ranges began to rise up from the plain.

Until the coming of plate tectonics theory, geologists had no real idea of how these mountains came to be. They realized that major mountain belts or 'orogens' are created in mountain-building events or 'orogenies' that last tens of millions of years then stop. They also realized that when orogeny ceases, erosion can wear mountains back to sea level in only a little longer.

Below: The world's great mountain ranges, with lofty, snow-capped peaks like these in the Rockies, Colorado, were thrown up where tectonic plates crunch together, crumpling and fracturing rock in the collision zone. Mountains like these are called fold mountains. But isolated high mountain peaks, like Kenya's Mount Kilimanjaro – some 5895m (19,340 ft) high – are created by volcanoes.

In 1899, the famous geologist William Morris Davis developed a beautifully simple life history for mountains, called the Cycle of Erosion. This envisaged the creation of mountains in a brief and violent spasm of uplift in the landscape then a gradual decline through 'youth', 'maturity' and 'old age' as forces of erosion such as rivers and the weather did their slow, steady work. Once the mountains were worn flat, another spasm of uplift started the cycle again. It did not explain how uplift took place, but Davis's model seemed so elegant that it was widely accepted – until the development of plate tectonics in the 1960s began to reveal an entirely new picture.

Tectonic mountain-building

Mountains can be built in a number of ways (see below-right), but it now seems clear that most great ranges form along plate boundaries.

The world's longest range is actually the mid-ocean ridge that forms where plates are moving apart under the sea.

It winds all through the Atlantic and up into the Indian Ocean. All the high ranges on land, however, occur where plates are moving together. Like a rug rumpled against a wall, converging plates crumple the rocks in between, forcing them upwards and creating long folds all along the boundary.

When an oceanic plate slides under a continental plate, offshore volcanic arcs and other debris are swept against the continent. Too buoyant to subduct, they become welded to the edge of the continent as an 'accreted terrane'. As the oceanic plate goes on pushing under the continent, these terranes pile up higher and higher in fractured and folded mountain belts. The North American Cordillera formed like this.

Eventually, all the oceanic plate may slide into the mantle, leaving the two continents to collide head-on. The enormous force of their collision crumples up their edges to build the greatest mountain ranges of all.

In the distant past this happened when Africa and North America collided to throw up the Appalachians,

Right: Mountains are also created where plates are pulling apart or 'rifting'. As the plates separate, magma wells up under the rift, stretching the crust and cracking it. Blocks of rock can then drop away down these 'faults' to open up a rift valley, flanked by mountains formed from blocks that didn't drop. The Basin and Range Province of the south-west USA (right) formed in this way.

and also when what are now called Europe and North America collided to throw up the Caledonian mountains of Scotland and Norway. Both ranges are now worn down to a fraction of their former height. Now the same process is happening as India ploughs into Asia to build the Himalayas (see below).

A more complex picture

In recent years, however, geologists have begun to realize that this basic scenario is only half the story. One complication comes from the discovery that crustal rocks are not simply rigid and brittle, but actually flow, albeit slowly. So the Himalayas, for instance, are more like a ship's bow wave in front of India than a rumpled carpet.

Moreover, other factors are involved in the process besides plate movement. When British scientist George Airy was surveying India in the 19th century, deviations of his plumbline revealed that the mass of the Himalayas extends far below the surface. In fact, we now know that all mountains have deep 'roots' that protrude far down into the mantle. This is because mountains, like

all the crust, float on the mantle. But because they are so big and heavy, mountains sink farther.

As mountain ranges are worn away by erosion, however, they get lighter and so actually float up. This phenomenon is called isostasy. When recent precise surveys showed the Appalachian Mountains are gaining a few centimetres in height each century, geologists were at first convinced the figures must be wrong, since the Appalachians are far from any plate boundary. In fact, it seems the Appalachians are rising isostatically – because erosion of rock in the valleys

so lightens them that the entire range is floating upward.

Erosion is slow, but it can wear mountain ranges flat in much the same time it takes to build them tectonically. So the process of uplift can actually be assisted by erosion.

Moreover, erosion can be sped up or slowed by changes in climate which can be altered by the way continents move, and even by mountains themselves. So geologists now realize that mountain building involves a complex interaction between erosion, climate, tectonic movements and isostatic adjustment.

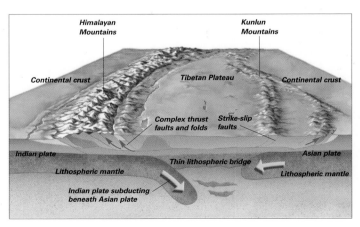

The Himalayas, the world's highest mountain range, began to form when the Indian plate crashed into southern Asia 55 million years ago. At the time Asia was made of softer, younger rock than India, and was powerless to resist the Indian advance. Long after the continents first crunched together, India has ploughed north at about 5cm/2in a year, pushing almost 2,000km/1,200 miles into the Asian plate. The Himalayas were thrown up as the crust doubled in thickness in the crumple zone. Because the Asian plate is warmer and lighter than the Indian plate, it is starting to ride over it, creating complex faults. But the creation of such high mountains interrupted the air flow over Asia, initiating India's famous monsoons. This intensifies erosion and is actually accelerating the ongoing uplift of the Himalayas as they are buoyed up isostatically by the weight of rock removed.*

EARTHQUAKES AND FAULTS

*The relentless movement of the Earth's tectonic plates can put the brittle rocks of the crust under such
stress that every now and then they crack altogether, and great blocks slide past each other along
fractures called faults, sending out shock waves that make the ground quake far around.*

There is an earthquake somewhere in
the world almost every day. Most are
so tiny they are only detectable on the
most sensitive of equipment. But a few
are big enough to cause devastation,
especially when they occur near major
cities, or trigger giant waves called
tsunami. Every year there are 20
quakes of the size that caused so much
damage in the Turkish town of Izmit in
1999. Giant quakes such as that which
triggered the tsunami that struck the
coastal regions of Southern Asia on
26 December 2004 occur once every
decade or so.

All kinds of things can trigger an
earthquake, from a landslide or a
volcanic eruption to the passing of a
heavy vehicle, and they can occur
almost anywhere. But most
earthquakes – and nearly all major
quakes – occur only in 'earthquake
zones' which coincide with the edges
of tectonic plates. In fact, an estimated
80 per cent of big quakes happen on
the edges of the plates around the
Pacific, and nearly all the rest happen
on the boundary between the Eurasian
and the Indo-Australian or Arabian
plates. The mid-ocean rifts often send
out tremors, but these are usually mild.

What causes earthquakes

Most quakes occur because of the
immense forces generated as two plates
grind past each other – either in
subduction zones where one plate dives
beneath another or along transforms
where two plates slide sideways past
each other. When one plate passes
another, the rock either side of the
crack may bend and stretch a little, but
sooner or later the stress builds up to
such a level that the rock suddenly
snaps. The sudden rupture sends shock
waves (seismic waves) shuddering out
through the ground in all directions
from the focus or hypocentre – the
point where the rock snaps.

*Above: Slicing right down through the state of
California, the San Andreas is one of the
world's most famous faults. Tremors set off by
bursts of movement along the fault shake the
cities of San Francisco and Los Angeles again
and again. Seismologists feel it is only a
matter of time before one of these cities is
struck again with a quake as big or bigger than
the one that devastated San Francisco in
1906. The San Andreas is not actually a single
crack but a series of strike-slip faults (see*

*Types of Fault, above right) that run along the
transform boundary between two tectonic
plates. To the west is the huge Pacific plate
which runs right under the Pacific Ocean. To
the east is the North American plate which
makes up most of the continent of North
America. Over the last 20 million years the
Pacific plate has moved 560km/350 miles
north – about 1cm/0.4in a year – but,
disturbingly, the pace seems to have
accelerated fivefold in the last century.*

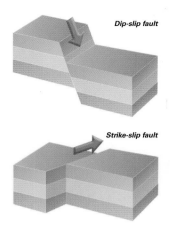

Dip-slip fault

Strike-slip fault

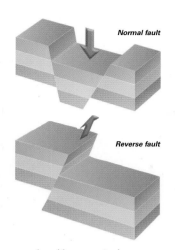

Normal fault

Reverse fault

Types of fault
Geologists classify faults by how the rock moves. A 'dip-slip' fault is one in which the rock slips up or down. A 'strike-slip' fault is one in which the rock moves sideways. Big strike-slip faults are called transcurrent faults, and typically lie along transform plate boundaries, where tectonic plates are sliding sideways. Dip-slip faults, on the other hand, occur where the crust is being squeezed or stretched horizontally. 'Normal' faults occur where tension pulls rocks apart, so that one block slips down. Rift valleys form where parallel sets of normal faults open up as the crust is stretched by upwelling magma. Where rock is squeezed – maybe by converging plates – a block may slide over another to create a 'reverse' fault. A thrust is a reverse fault that slides up at a shallow angle.

The rupture spreads along the plate boundary like a crack spreading through glass. The longer the crack, the bigger the quake.

In the southern Asia quake of 2004, the rupture ripped 1,000km/620 miles along the Indo-Australian plate boundary. The massive Alaskan earthquake of 1964 moved entire mountains up 12m/40ft.

Most earthquakes only move the ground a few centimetres or so. Yet the cumulative effect of successive earthquakes has a bigger impact on the landscape. If the rocks either side of the crack or fault move just 10cm/4in a century, over a million years they can move up or down 1km/0.6 miles.

Earthquake waves
There is no chance of running from an earthquake. There are four kinds of shock wave – P or Primary and S or Secondary waves underground, and Love and Rayleigh waves on the surface. All move very fast. P-waves, which shake the ground up and down, are the fastest, roaring along at 5km/ 3 miles per second. S-waves, which snake from side to side, travel only a little slower. Surface waves are slower but these are what do the damage when an earthquake strikes cities. In solid rock, the waves move too fast for the eye to see. But they can turn loose sediments – in, for example, vulnerable areas such as landfill sites – into fluids so that waves can be seen rippling

across them like waves in the sea. These waves can capsize buildings. In both the Kobe (Japan) quake of 1995, and the San Francisco tremor of 1989, the worst damage was to buildings built on landfill sites.

Quake danger
Many of the world's cities – Los Angeles, Mexico City, Tokyo – are sitting on timebombs, because they are right in the middle of earthquake zones. People in these cities have learned to live with minor tremors. But sooner or later, one may be hit by a devastating quake that does terrible damage. For these people, learning to predict a quake is a race against time.

Below: The quake which devastated Marmara and Izmit in Turkey in 1999 seems to be one of a series moving west along the North Anatolian Fault and may reach the capital, just 50 miles (80km) away.

One approach is to look at the historical record. If there has not been a quake in an earthquake zone for some time, the chances are there will be one soon. The longer it has been quiet, the bigger the quake will be, as strain has been building in the rocks.

Most seismologists believe the key is to watch for strain building in rocks. In many earthquake zones, high-precision surveys monitor the ground for signs of deformation. Laser ranging from satellites such as the Japanese Keystone system can make acutely fine measurements. There is now some evidence that earthquakes occur in clusters, as one quake sets up stress further along a fault that will be in turn released as a quake. So a moving succession of quakes may be triggered off, as seems to be happening westwards along the North Anatolian Fault in Turkey, towards Istanbul.

VOLCANOES

Few sights are more awesome than a volcano erupting in a huge explosion of gas, ash and lava, and their effect on the landscape is immediate and dramatic. Volcanic activity is also going on beneath the Earth's surface. The combination of all this vulcanicity has a profound effect on the Earth's geology.

Volcanoes are places where red-hot magma (molten rock) wells up through the Earth's crust and erupts on to the surface. They are not randomly located around the world but clustered in certain places – where there is a ready supply of magma. Despite the inferno of heat in the Earth's interior, immense pressure keeps most rock in the mantle beneath the surface solid. But along the margins between the great tectonic plates that make up the Earth's surface, mantle rock melts into magma in huge volumes, and buoyed by its relatively low density wells up to the surface to erupt as volcanoes.

All but a few of the world's active volcanoes lie close to plate margins – mostly where plates are converging – and especially in a ring around the Pacific Ocean known as the 'Ring of Fire'. The exceptions are so-called 'hot-spot' volcanoes, such as Mauna Loa in Hawaii, which well up over fountain-like concentrations of magma called mantle plumes.

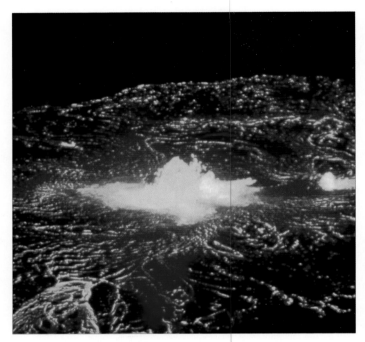

Above: Lava is the name for magma that has erupted on to the surface. It is so hot – over 1,100°C/2,000°F – the rock flows like a river.

Volcanic eruptions vary widely in character. Along the ocean bed ridges where plates are moving apart – and over hotspots – runny, silica-poor 'basaltic' magma wells up through cracks that ooze red-hot lava almost continuously in gentle spouts. Some volcanoes belch ash and steam. Others eject showers of pulverized rock. Some unleash devastating mudflows as the heat melts ice, and some spew glowing avalanches of cinders and hot gas.

Explosive eruptions

The most terrifying and least predictable volcanoes of all tend to be those along the margins where plates are crunching together during subduction. Here, magma melting its way up through the thick plate margins becomes so contaminated with silica that it becomes 'dacitic' – so

thick and viscous that it frequently clogs up the volcano vent. It seems as if the volcano is sleeping or even dead – until, almost without warning, the pent-up pressure bursts through the plug in a cataclysmic explosion that hurls out shattered fragments of the plug (called pyroclasts), huge jets of steam and clouds of ash and cinder, as well as streams of lava.

The driving force in explosive eruptions is the boiling off of carbon dioxide gas and steam trapped within the reservoir of magma beneath the volcano called the magma chamber. The more gas and water present in the magma, the more explosive a volcano becomes. Magma near subduction

Below: Most explosive eruptions begin with the blasting of a massive cloud of gas, ash and steam high into the atmosphere.

zones often contains ten times as much gas as elsewhere. Gas in magma can expand to hundreds of times the volume of molten rock in a matter of seconds.

Types of eruption

No two volcanos are quite the same, but vulcanologists identify a number of distinctive styles of eruption. 'Effusive' eruptions occur when fluid basaltic lavas ooze from fissures and vents and flood far out over the landscape to form a plateau or a shallow dome. In some, lakes of lava pool up around the vent, while others shoot sprays of lava into the air. These sprays or fire fountains are driven by bubbles of gas in the lava, just like the droplets sprayed from a fizzy drink.

The explosive eruptions that occur with more viscous magmas are much more variable in character. Among the mildest are 'Strombolian' eruptions named after the island of Stromboli off the west coast of Italy. Gas escapes sporadically and the volcano repeatedly spits out sizzling clots of lava, but there is rarely a really violent explosion. 'Vulcanian' eruptions, named after Vulcano in the Italian Lipari islands, are much more ferocious. Here the magma is so viscous that the vent frequently clogs, in between roaring, cannon-like blasts that eject ash-clouds and fragments of magma followed by thick lava flows.

'Pelean' eruptions blast out glowing clouds of gas and ash called *nuées ardentes*, such as Mount Pelée, Martinique, in 1902. 'Plinian' eruptions are the most explosive of all, named after the Roman author Pliny, who witnessed the eruption of Vesuvius that buried Pompeii. Boiling gases blast clouds of ash and volcanic fragments high into the stratosphere.

Right: Hawaiian volcanoes like Kilauea – one of the most active on Earth – are famous for their spectacular jet-like sprays of liquid lava called fire fountains. They can occur in short spurts or last for hours on end. Occasionally they can shoot hundreds of yards into the air, and in 1958 one shot up to almost 610m/2,000ft. However, this was dwarfed by a fire fountain on the Island of Oshima, Japan, in 1986, which reached 1,524m/5,000ft!

Types of volcano

Volcanoes can also be classified by the kind of cone they create. The instantly recognizable cone-shaped volcanoes like Japan's Mt Fuji are composite or 'stratovolcanoes' created where sticky magma erupts explosively from a single vent. Successive eruptions build the cone from alternate layers of lava and the ash that rains down on it. Some cones, though, are built entirely of cinders and ash, such as Mexico's Paricutin. Shield volcanoes such as Hawaii's Mauna Loa are shallow, dome-shaped volcanoes formed where fluid lava spreads far out from a single vent. Fissure volcanoes are ridges created when runny lava oozes from a long crack. Large scale fissures occur along the mid-ocean ridges. Smaller ones burst through on the flanks of larger volcanoes.

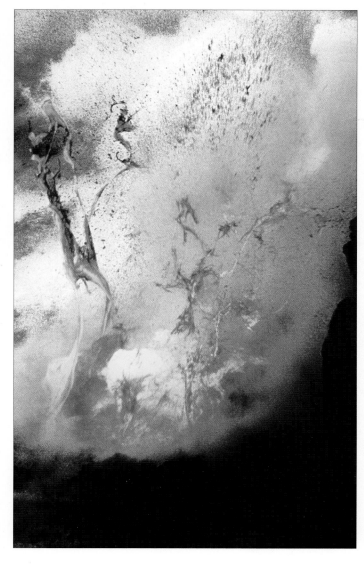

IGNEOUS FEATURES

Although eruptions are brief, volcanic activity leaves a lasting legacy as magma, lava and volcanic ash harden to form rock. Volcanic peaks and ash deposits are clearly visible on the surface, but much of the molten magma that wells up from the Earth's interior remains trapped underground, solidifying in situ.

Rock features formed by volcanoes above ground and magma underground are together termed igneous features. Features formed from magma trapped underground are termed intrusive igneous features. Those formed above ground are called extrusive features.

Intrusions underlie all the world's major continents, and in many places they have been exposed on the surface after erosion of the surrounding 'country rock'. Intrusive igneous rock, which includes granite and gabbro, is almost invariably very tough and crystalline, and endures the weather to stand proud long after the surrounding softer rocks have been worn away.

Massive intrusions
As seen in Types of Intrusion, opposite, the large intrusions forming deep underground are called plutons, and typically made of granite. Sometimes scores of plutons can coalesce over time to form monsters called batholiths (from the Greek for 'deep stone'), which lie like giant whale carcasses under most of the world's great mountain ranges. North America's Coast Range batholith extends 1,500km/932 miles under British Columbia and Washington. Batholiths mostly form as the convergence of two of the world's great tectonic plates generates huge quantities of magma underground. Sometimes, batholiths are topped by smaller protuberances called stocks and bosses. Erosion may expose several bosses on the surface, as at Dartmoor and Bodmin Moor in south-west England, which are linked underground to a single batholith.

Minor intrusions
Nearer the surface, smaller intrusions often occur as sheetlike formations called dykes and sills. Dykes are anything from a few centimetres to

hundreds of metres thick, and form as magma is injected into cracks in the rock. They are typically near vertical, but any sheet intrusion that cuts right across layers of country rock may be described as a dyke. Because they cut across existing structures, dykes are described as 'discordant'. Often, dykes form as the pressure of upwelling magma fractures overlying rock, opening cracks that fill with magma. A single intrusion may breed dozens of dykes like this in a 'dyke swarm'. Occasionally, these form in concentric downward-pointing cones or cone sheets crowning the intrusion, as famously on Mull in Scotland.

Ring dykes form like the sides of an upturned pan or cauldron as a round block of country rock drops away,

Above: In California's Yosemite Park, spectacular cliffs of grey rock rear up where erosion has stripped naked the ironhard granite batholith that underlies much of the Sierra Nevada mountains. The Half Dome, seen above, is the most impressive example.

leaving a gap to fill with magma. Classic ring dykes like this are seen in Glencoe in Scotland and Mount Holmes in Yellowstone Park.

Sills are typically horizontal or gently sloping sheets, forming as magma seeps between existing bedding planes. Because they follow existing structures, they are described as 'concordant' structures. Sometimes the magma arches up between rock beds to form dome-shaped laccoliths or warps the rock downward to form dish-shaped lopoliths.

Massive extrusions

Geologists have long debated over how vast intrusions form (see The Granite Problem under Granite, World Directory of Rocks), but extrusions did not seem to pose the same problems, because volcanoes on land produce only small volumes of lava. Yet not only is the entire ocean floor made of extrusions of basalt erupted from mid-ocean fissures, but in places around the world there are solidified remains of gigantic lava flows.

Flood basalts are the solidified remains of huge floods of basalt lava that erupted in the past. The most spectacular is the Deccan Traps of India. Over 2km/1.2 miles thick and covering 500,000sq km/200,000sq miles, the Deccan Traps include half a million cubic km/120,000 cubic miles of lava – half a million times as much as erupted at Washington's Mount St Helens, USA, in 1980! The Traps' basalts erupted about 65 million years ago in a gigantic outpouring that has been blamed by some palaeontologists for the death of the dinosaurs.

Another huge flood basalt is the Columbia River plateau of north-western USA, which erupted 175,000 cubic km/42,000 cubic miles of lava about 16 million years ago. Others include southern Africa's Karoo and the Faroe Islands, North Atlantic.

Undersea floods

These flood basalts were originally thought to be rare exceptions. Then in the late 1980s, seismic surveys of the sea floor began to reveal even more spectacular examples of what came to be called Large Igneous Provinces or LIPs under the oceans.

The largest of these, called the Ontong Java Plateau, lies under the Pacific to the east of Borneo and covers nearly 5 million sq km/2 million sq miles, an area bigger than the whole of the USA. This enormous Igneous Province erupted less than three million years ago.

Above: India's Deccan Traps were the site of one of the greatest eruptions of lava in the Earth's history, about 65 million years ago. Traps is Dutch for staircase, referring to the steplike shape of the eroded layers of lava.

Geologists believe 'mantle plumes' cause these massive eruptions. These are fountains of hot matter that rise through the Earth's mantle, perhaps all the way from the D" layer on the core boundary (see Inside the Earth, this section). The theory is that as they rise, the resulting heat melts mantle rock to create huge quantities of molten magma which burns through the crust and floods on to the surface.

Types of intrusion

Intrusions vary widely in size and shape. Deep underground, diapirs (rising blobs of magma) open up huge spaces in the country rock to form lumps of igneous rock called plutons. These include drum-shaped stocks a few kilometres across and gigantic batholiths made from scores of plutons and stretching thousands of kilometres. Nearer the surface, magma intrudes into cracks in the rock to create igneous rocks in thin sheets (dykes and sills) or lens shapes (lopoliths and laccoliths). Where these minor intrusions cut across existing structures, as dykes often do, they are said to be discordant; where they follow existing structures, like sills, they are said to be concordant. Minor near-surface intrusions are often exposed on the surface by subsequent erosion of the surrounding country rock, but even deep-forming plutons may be exposed this way over a considerable period of time.

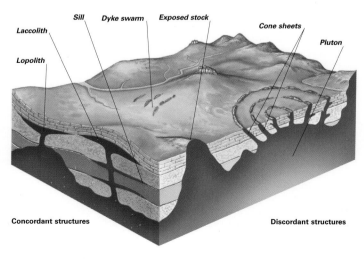

THE ROCK CYCLE

Mountains and hills look so solid, it's hard to imagine they could ever change. Yet all the Earth's landscapes are being continually remodelled as rocks are attacked by the weather and worn away by running water, moving ice, waves, wind, and other 'agents of erosion'.

Occasionally, the landscape is reshaped suddenly and dramatically – by an avalanche or landslide. Yet most of the time it is remoulded slowly but relentlessly. One cold night does little damage. A single shower of rain seems to run straight off. Yet night after freezing night, shower after shower, repeated over millions of years, takes its toll. Research into rates of erosion has shown it to average at least 0.5mm/0.04 inches a year on land. At that rate, even a mountain range as high as the Himalayas can be worn entirely flat in just over 20 millions of years.

Some new landforms are created from old as rock is denuded or stripped away by weathering and erosion. Some are created by steady accumulation or deposition of the rock debris. Most hills and valleys are formed by a combination of both denudation and deposition.

Below. On coasts, shorelines are battered by waves packed with energy by wind blowing far over the ocean. Waves hurl tons of water filled with shingle against coastal rocks and ram air into cracks so that rocks are forced apart.

Weathering

As soon as rock is exposed to the weather, it gradually starts to break down under the assault of wind and rain, frost and sun. Sometimes the rock is corroded by chemical reactions caused by moisture in the air, or water trickling over it. Sometimes they are attacked by micro-organisms and lichens or chemicals released by plants. Sometimes they are broken down physically by, for instance, the effects of heat and cold. Water in cracks can expand so forcefully as it freezes that it can shatter rock. At -22°C/-7.6°F, ice can exert a pressure of 3,000kg/6,600lb on an area the size of a coin. This is called frost-shattering.

Geologists argue about the relative importance of chemical and physical weathering. Some limestone regions – known as 'karst' – show all the signs of chemical weathering which opens up potholes and spectacular caverns (see Rock Landscapes in this section). Other regions show strong signs of physical weathering. Frost-shattering

creates the jagged peaks in high mountain regions – as well as the piles of debris called scree below. In most places both chemical and physical processes are at work.

Erosion and deposition

All the debris created by weathering is gradually carried away by agents of erosion – the most important of which is water. Without running water to mould it, the landscape would be as jagged as the surface of the Moon. Rivers and streams slowly soften contours – wearing away material here, depositing it there. Over millions of years, a river can carve a deep canyon or spread out a vast plain of sediment as it flows towards the sea.

In deserts, however, running water is scarce. Intermittent streams carve out valleys, but the landscape is angular, and some landforms are carved into weird shapes by the blast of windblown sand. On coasts, it is the waves which are the dominant agents of erosion, and their continuous

Right: In many parts of the world, the landscape is shaped by water running over the land. Rivers can carve out deep valleys as they run down to the sea. A river's steady torrent washes small grains loose and grinds away solid rock with the stones it carries. Arizona's Grand Canyon (pictured here) was carved out as the Colorado river cut its way through a rising landscape over many millions of years.

pounding on the shore creates a distinctive range of coastal landforms, including steep cliffs where waves cut into hillsides, platforms of rock sliced out as the waves wear back into the cliff and stacks of rock left behind as the waves erode the cliff.

It is far from certain how much of today's landscape has been created by the processes we see operating today – and how much by more dramatic events in the past. Some geologists argue that some river-eroded features in deserts were carved out during especially wet periods in the past called pluvials. There is also no doubt that cold periods in the past known as ice ages were crucial in shaping the land in Europe and North America. Giant U-shaped valleys in the mountains of north-west USA and Scotland, fjords in Norway and Canada and vast deposits of 'till' (rock debris) covering the USA's Midwest can only have been created by moving ice.

Rocks remade

Just as erosion relentlessly destroys rock wherever it is exposed on the surface, so new rocks are created all the time. These new rocks are often forged from remnants of the old, so that rock is continually recycled in a process called the rock cycle (below).

Not all material moves through the cycle at the same rate. Some material in rocks in continental interiors may sit virtually unchanged for billions of years, while material in rocks near the active margins of continents – those

near coasts and near to subduction zones – has been through the mill again and again.

Some material eroded from rocks is taken back down into the mantle on subducted sea-beds. So too is rock forming oceanic plates. Some may melt as it descends and rise with magmas to form new igneous rocks. Some will be carried right down into mantle. Even there it is not necessarily lost forever. Convection makes material circulate right through the mantle, so it may re-emerge, eventually, as molten magma – even if the process takes hundreds of millions of years. Studies have shown that most of the atoms in rocks in the continental crust came up from the mantle over 2.5 billion years ago. Yet the rocks they are in are usually much, much younger, since these atoms have been recycled many times.

Rock breaking

New rock material is sometimes brought up from the mantle in magmas, but most surface rocks are made from ingredients that are continually recycled as rocks are made and remade. The ingredients can be large outcrops of rock, or small chunks, grains or even atoms, and geologists call the recycling process the rock cycle. There are many paths through the cycle. Igneous rock formed by the freezing of molten magma, for instance, might be broken up by the weather into fragments that are washed in rivers into the sea. There fragments pile up on the sea bed and eventually turn to sedimentary stone. This sedimentary rock may, in turn, be buried and squeezed or heated to form metamorphic rock. This too can be broken down and made into sedimentary rock.

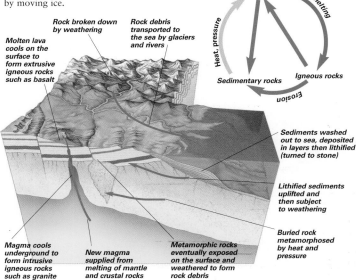

Rock broken down by weathering

Rock debris transported to the sea by glaciers and rivers

Molten lava cools on the surface to form extrusive igneous rocks such as basalt

Metamorphic rocks

Melting

Heat, pressure

Sedimentary rocks

Igneous rocks

Erosion

Sediments washed out to sea, deposited in layers then lithified (turned to stone)

Lithified sediments uplifted and then subject to weathering

Buried rock metamorphosed by heat and pressure

Magma cools underground to form intrusive igneous rocks such as granite

New magma supplied from melting of mantle and crustal rocks

Metamorphic rocks eventually exposed on the surface and weathered to form rock debris

HOW IGNEOUS ROCKS FORM

No rocks undergo quite such a change as they form as igneous rocks, the rocks that form almost all the ocean crust and a great deal of the continental crust. Before they solidified into their current form, these extraordinarily tough, crystalline rocks were glowing, searing hot liquid magma.

Igneous rocks are quite literally frozen magma, hot molten rock from the Earth's interior. Magma may be quite thick and sticky, but it is quite genuinely a fluid and flows like one. The process that turns it to solid rock is exactly the same that turns water to ice when the temperature drops to 0°C/32°F. The difference is that magma freezes at much higher temperatures than water – anywhere between 650°C/1,200°F and 1,100°C/2,000°F. Magma also contains a complicated mix of substances, each with its own freezing point, so magma does not freeze in one go like water but bit by bit.

In magma, elements such as silicon, iron, sodium, potassium, magnesium and so on occur in pure form, or as simple compounds. Yet as the magma cools, these elements and compounds join to form crystals of various minerals. The commonest minerals to form are quartz, feldspars, micas, amphiboles, pyroxenes and olivine. Magma also contains gases such as water vapour, sulphur dioxide and carbon dioxide, but these are driven off during cooling.

As magma solidifies, molecules which were vibrating wildly when it was liquid begin to calm down enough to form clusters. These clusters soon begin to grow and form crystals here and there in the melt. They grow especially quickly near the surfaces, which cool fastest. This is why a solid crust can form on the surface of lava which is hot and fluid on the inside.

Coarse and fine

The slower magma cools, the larger crystals grow. In intrusions below ground, cooling can take thousands, if not millions, of years, and crystals grow big enough to see with the naked eye. Rocks with such crystals are said to be phaneritic, or coarse-grained.

Cooling is usually much quicker in lavas – that is, magmas which erupt on the surface to form extrusive rocks. Cooling takes a matter of weeks or even days. Crystals have little time to grow and so are often only visible under a microscope. Rocks with such crystals are said to be aphanitic, or fine-grained. Sometimes a fine-grained rock may contain larger crystals that formed earlier, while the magma was still underground. These large crystals are called phenocrysts and the rock is said to be porphyritic.

Where lava is ejected in small blobs, it can cool in a matter of hours – so quickly that crystals cannot form at all. The result is a glass, like obsidian, in which there are no crystals at all.

How igneous rocks evolve

Each mineral crystallizes at a different temperature. Olivine and pyroxene set at over 1,000°C/1,832°F. Silicate minerals such as quartz don't freeze until temperatures are as low as 650°C/1,200°F. So magmas crystallize progressively, with some minerals forming earlier than others.

In the 1920s, laboratory tests conducted by Norman Bowen showed how minerals crystallize in a sequence (Bowen's Reaction Series), from high to low temperature: olivine, pyroxene, amphibole, biotite mica, quartz,

Left: There can be no more dramatic demonstration of the toughness of granite than these peaks in Torres del Paine, Chile. The softer country rock into which the granite magma was intruded has been stripped away by millions of years of erosion.

Above: Granite magmas cool slowly underground and crystals grow large enough to be seen with the naked eye, as in this specimen of Cornish granite. The white crystals are quartz, the black ones biotite mica and the pink ones potassium feldspar.

muscovite mica, potassium-feldspar and plagioclase feldspar.

This sequence shows how igneous rocks form and how each kind develops its own mineral make-up, according to the circumstances in which it forms. It also provides a mechanism for igneous rocks to evolve. Without such a mechanism, the granitic magmas that form our continents could never have developed.

The idea is that when it formed, the Earth was much like the Moon, made mostly of just a simple parent rock – a 'mafic' or 'ultramafic' rock low in

silica like basalt and peridotite, but unlike granite (which is silica-rich). From this basic start all the many other kinds of igneous rocks evolved by a process called fractionation. This is the way the composition changes during either melting or freezing of the rock, as different minerals melt or freeze first.

When a mafic rock melts, for instance, it splits into two fractions as low temperature minerals such as quartz and feldspar melt first and flow away, leaving high temperature minerals behind. Low temperature minerals are all silicates. So the melt is much richer in silicate minerals than the original rock. Successive refreezing and melting further boosts the silica-content to create silica-rich granite. Which of the many kinds of igneous rock that ultimately forms depends on a variety of factors, including the depth at which the magma is generated and its history of fractionation.

Where igneous rocks are found

Igneous rocks form only in certain places. Fractionation occurs mostly where tectonic plates are either moving apart at mid-ocean ridges, or pushing together at subduction zones. At mid-ocean ridges, fractionation of the parent magma from the mantle creates

Above: Clearly visible under a microscope in this fine-grained igneous rock are larger crystals or phenocrysts that crystallized slowly underground before the lava erupted to the surface.

basalt lava at the surface and gabbro deeper down. At subduction zones, partial melting of the subducted plate fractionates to create intermediate rocks such as diorite in island arcs. Further melting and remelting, especially beneath continents, creates granite. Granite only forms underground, but it can also melt to create the silica-rich lava rhyolite.

Below: All igneous rocks form when red-hot molten magma like this freezes, either on the surface or underground. The temperature at which the magma freezes, and the kind of rock that forms, depends on the balance of the various chemical elements in the magma.

HOW SEDIMENTARY ROCKS FORM

Over 90 per cent of the Earth's crust is made from igneous rock, but on land, 75 per cent of it is hidden beneath a veneer of sedimentary rock, rock formed from sediments laid down in places such as the sea-bed, buried and turned into stone over millions of years.

Sedimentary rocks start to form wherever sediments settle on the beds of oceans, lakes and rivers, or are piled up by moving sheets of ice or the wind in the desert. As sediments build up, layers are buried ever deeper, drying and hardening as water is squeezed out. Over millions of years, the pressure of overlying layers and the heat of the Earth's interior turns layers of sediment to solid rock in a process called lithifaction.

When sediments are powdery and soft and contain few hard sand grains, compaction alone is enough to turn them to stone. Very sandy sediments, however, are too hard to be compacted so easily. To turn to stone, sandy sediments must be glued together by cements made from materials dissolved in the water from which the debris settled. The most common cements are silicate minerals, calcite and iron compounds which give rocks a rusty red look.

Beds and joints

Because sediments settle layer upon layer, outcrops of sedimentary rocks are usually marked by a distinctive layered or 'stratified' look. Where the rock has been undisturbed since turning into stone, these layers appear horizontal. You are not very likely to

Above: This ammonite fossil, preserved in chalk, once inhabited shallow waters.

see such uniform layering, however, as the movements of the Earth's crust twists them into contorted shapes.

The thinnest layers or beds are marked out by lines called 'bedding planes'. Beds may be the sediments laid down in just a single season. Yet the thickest layers or strata may have taken millions of years to build up. Sedimentary rocks may also be marked by 'joints', cracks across layers formed as the rock dried out and shrank.

Sedimentary rocks usually contain another distinctive feature – fossils, the remains of living things turned to

Enduring sand

Of the three major ingredients of clastic rocks, quartz is by far the most durable, surviving the destruction of its parent rock again and again. Once freed from igneous rocks by weathering, quartz accumulates in layers. Grains are then cemented together into sandstones (right) like the de Chelly beds of the buttes in Utah's famous Monument Valley (below). Tough though these rocks are, they too are broken down in time. Yet it is not the sand that is destroyed but just the cement. The liberated sand now scattered on the desert floor will, in time, form new sandstones.

Right: Sedimentation is very rarely continuous, but occurs in fits and starts. The result is that many sedimentary rocks, like these sandstones in Utah's famous Zion Canyon, are marked by countless bedding planes marking a brief pause in sedimentation.

stone. Fossils help geologists to determine the conditions in which certain rocks formed. They can tell chalks formed in shallow tropical seas, for instance, because they are studded with fossils of sea creatures that live only in such conditions. Moreover, different creatures and plants lived at various times in Earth's history. So geologists can also work out the relative ages of sedimentary rocks from the range of fossils they contain, a process known as biostratigraphy.

Kinds of sedimentary rock

Indeed, some sedimentary rocks, like limestone, are made almost entirely from the remains of living organisms, or else by chemicals created by them. Such rocks are described as organic. Chemical sediments such as evaporites are made from minerals precipitated directly from water. Most sedimentary rocks, however, are 'clastic' or 'detrital' which means they are made from fragments or clasts of rock broken down by the weather.

Rocks from debris

When rocks such as granite are weathered, they eventually crumble to form clasts. The different minerals crumble in different ways. Quartz crystals, for instance, are so hard they are left as distinct sand grains, while orthoclase feldspar breaks down into clay and plagioclase forms calcite. Although this debris all starts off together, the different kinds of clast are gradually separated as they are washed down by rivers into seas and lakes.

Moving water is like a natural sieve, sorting the rock fragments into small and large by carrying smaller grains farther and faster. The farther from the source they are carried, the more they

Right: Few rocks are so distinctive as white chalk, seen here in cliffs at Normandy, France. It is made of calcite-rich remains of countless marine plankton that floated in ancient seas.

become sorted. So typically only rocks that form fairly near the source, such as some conglomerates, contain a full range of particle or grain sizes. Most rocks contain particles predominantly of a particular size of grain.

The result is that clastic sedimentary rocks can be divided into three groups according to grain size: large-grained 'rudites' such as conglomerates and breccias; medium-grained 'arenites' such as sandstones; and fine-grained 'lutites' such as shale and clay.

As a river washes into the sea, or into a lake, the heaviest quartz grains are dropped nearest the shore, forming sandstones. Finer clay particles are

washed farther out, forming shales. Dissolved calcite is washed farther out still, and only finally settles to form limestone rocks. This only happens once the particles are taken from the water by living things which use them for building shells and bones. When they die, they take the calcite down to the sea floor in their remains.

In some rocks, such as wackes, there is a mix of grain sizes. However, even wackes may be banded into layers, with a gradation of grain size from fine at the top to coarse at the bottom. This 'graded bedding' develops because the largest, heaviest grains settle out of the water first.

HOW METAMORPHIC ROCKS FORM

When rocks are seared by the heat of molten magma or crushed by the enormous forces involved in the movement of tectonic plates, they can be altered beyond recognition. The crystals they are made from re-form so completely that they become new rocks, called metamorphic rocks.

Early geologists soon appreciated that igneous rocks formed from melts and sedimentary rocks from sediments. Yet there was a third very common kind which they could not quite pin down.

Slate used for roofs, for instance, has a colour and texture like shale. Yet it is much harder than shale and splits into sheets that are completely unrelated to bedding. An even harder rock called gneiss found in the Alps has strange swirls and bands. Beautiful white marble seems to be made of calcite like limestone, yet contains no fossils and has a dense crystalline structure not unlike granite.

Marble seemed to represent a cross between sedimentary and igneous rock and geologists began to realize that

this could provide the clue to its origin. Marble was indeed once limestone but it has been altered by heat. The calcite it is made of has been cooked and its crystals reformed in a way that looks more like an igneous rock. Marble, like many other rocks, is metamorphosed or re-formed by heat, or pressure, or both. The word metamorphic comes from the Greek for 'change form'.

Cooking and crushing

It is now clear that just about any rock – igneous, sedimentary and metamorphic – can be metamorphosed into new rock. To create a metamorphic rock, the heat and pressure were extreme enough to

completely alter the original rock or 'protolith', but not so extreme as to melt it or break it down altogether. Heat intense enough to melt it would have created an igneous rock.

Rocks are subjected to heat and pressure either by being buried deep in the crust, or being crushed beneath converging tectonic plates. They are subjected to heat alone by proximity to hot magma.

Heat and pressure metamorphoses rocks in two ways. First, it changes the mineral content, by making minerals react together to form new ones. Some minerals are unique to metamorphic rock. When shale is metamorphosed to slate, for instance, its clay is changed to chlorite, a mineral only found in metamorphic rock.

Second, it changes the size, shape and alignment of crystals, breaking down old crystals and forming new ones in a process called recrystallization. The original rock might be made of just a single mineral. When this is metamorphosed the mineral recrystallizes in a different form. Pure quartz sandstone becomes quartzite. Pure calcite limestone becomes marble.

Metamorphic environments

Each kind of metamorphic rock has its own particular protolith or set of protoliths. Marble is only formed from pure calcite limestone. However, mylonite can form from practically any rock. Each kind of rock also forms only in specific conditions. Mild metamorphism turns shale to slate. Yet if the heat and pressure become more intense, it turns first to phyllite, then to schist and finally to gneiss.

Left: Like so much of the Canadian shield, the mountains of Churchill are made from ancient, very tough metamorphic rock. This quartzite landscape formed when sandstones were metamorphosed by intense heat and pressure, some two billion years ago.

It has become clear that particular combinations of conditions are likely to create particular metamorphic rocks. So geologists talk about environments of metamorphism. They also focus on sets of conditions that form particular combinations or facies of metamorphic minerals (see Metamorphic facies under Schists, World Directory of Rocks). One of the most important environments is next to hot igneous intrusions. An intrusion may have a temperature of 900°C/1652°F, and literally cooks the rock it comes into contact with. This is called 'contact metamorphism' and involves heat alone without significant pressure.

In fault zones, rock is ripped apart by tectonic movements. Near the surface, this shatters rock to form a breccia or crushes it to a powder. Deeper down, rock is so warm that it smears out rather than breaks. As it does it is subjected to 'dynamic metamorphism' which creates the rock mylonite.

Regional metamorphism

The huge forces involved when continents collide and throw up mountain ranges can crush and cook rocks over large areas. Near the fringes, this regional metamorphism can be mild, and mudrocks are altered to low-grade metamorphic rocks such as slate and phyllite. Towards the heart of the mountain belt heat and pressure gradually increase. Under moderate heat and pressure, slate and phyllite are metamorphosed to schist. Intense heat and pressure (high-grade metamorphism) creates gneiss. Even more extreme conditions can partially melt the rock to create migmatite.

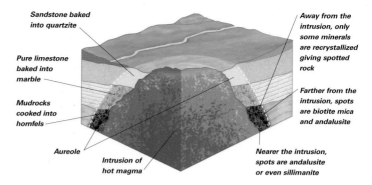

Sandstone baked into quartzite

Pure limestone baked into marble

Mudrocks cooked into hornfels

Aureole

Intrusion of hot magma

Away from the intrusion, only some minerals are recrystallized giving spotted rock

Farther from the intrusion, spots are biotite mica and andalusite

Nearer the intrusion, spots are andalusite or even sillimanite

Metamorphic grading

The most widespread metamorphic environment is where tectonic plates converge with the force to throw up mountains. Beneath the mountains, rocks are crushed and sheared, and heated by rising magma and their proximity to the Earth's mantle. The scale of this is enormous and is called 'regional metamorphism'. The intensity of regional metamorphism is described in terms of grades. Low-grade metamorphism means low temperature (below 320°C/608°F) and pressure. High-grade metamorphism means high temperature (above 500°C/932°F) and pressure. These different grades of metamorphism create different minerals and different structures.

Metamorphic rocks that form where plates converge have a very distinctive characteristic. Squeezed between the plates, new crystals are forced to grow flat, at right angles to the pressure. The

Contact metamorphism

When rocks are cooked by the heat of an intrusion, the result is contact metamorphism. Around the intrusion is a ring or 'aureole' of affected rock. The way in which particular rocks are affected depends on how close they are to the intrusion and the intrusion's size.

crystals are so intensely aligned that the rocks have layered structure, like the leaves of a book and so called foliation. Foliation means slate breaks easily into flat sheets, it gives schist a stripey look, called schistosity, and gneiss even more dramatic swirling bands, as the minerals are separated out into layers. Foliation is such a distinctive characteristic of regionally metamorphosed rocks that all metamorphic rocks – not just regionally metamorphosed rocks – are divided into foliated rocks (such as slate, phyllite, schist and gneiss) and non-foliated rocks (such as hornfels, quartzite and amphibolite).

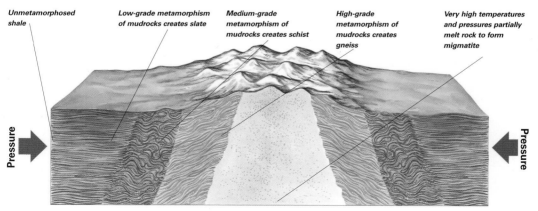

Unmetamorphosed shale

Low-grade metamorphism of mudrocks creates slate

Medium-grade metamorphism of mudrocks creates schist

High-grade metamorphism of mudrocks creates gneiss

Very high temperatures and pressures partially melt rock to form migmatite

Pressure

Pressure

ROCK LANDSCAPES

Every kind of rock and geological formation produces its own distinctive kind of landscape, from the jagged ice-capped peaks of young fold mountains and the spectacular gorges and caverns of some limestone regions to the gently rolling landscapes of chalk downland.

The relationship between rocks and landscapes is a complex one, and geologists have been trying to unravel it for centuries. Yet an experienced geologist can tell a great deal about the nature of the rocks and the rock formations simply by studying the shape of the land.

Some landforms give an instant clue to the kind of rock involved. Gorges with pale rock faces, spectacular caverns and deep potholes cannot be anything but limestone (see opposite page). Granite tors are also easy to spot. Similarly, long mountain ranges with soaring, jagged peaks are clearly fold mountains, created by the convergence of two tectonic plates.

Hills and vales

Most of the time, however, the relationship is more subtle. On the whole, the hardest rocks, such as granites, gneisses, sandstones and limestones tend to resist erosion and form hills. Softer rocks like clays and

Above: Spitzkoppe in Namibia is one of many steep-sided granite monoliths in the tropics called inselbergs. It was once thought only deep weathering in ancient climatic conditions could have produced such outcrops. Now it is believed that the shape simply reflects the way ordinary weathering attacks the structure of the original intrusion.

Below: Salt Cellar Tor in England's Peak District is one of several tors in this region. They are made of the tough sandstone and shale rock series millstone grit. These tors probably formed when rock was weathered underground by water seeping into joints – perhaps during the ice ages. Weathered rock was then stripped away to reveal the tor.

mudstones are worn away into valleys. There can also be valleys in hard rock, and tectonic movements can uplift soft rock to create hills or even mountains. Clay very rarely forms hills, however, because it is so quickly worn away.

In south-east England, and places in the Appalachians, gentle folding has tilted layers of sedimentary rock to create a distinctive 'belted' landscape of parallel ridges and valleys. Layers of softer rock, such as clay, are eroded faster to create valleys, while the harder rocks such as sandstones and chalk resist erosion to form ridges.

One side of the ridge is a steep 'scarp' slope where erosion has cut right across the strata. The other side, however, slopes gently down the top surface of the rock layer and is called the 'dip' slope. A ridge shaped like this, with one steep scarp and a gentle

dip slope, is called a cuesta. The angle of the dip slope gives an instant clue to the angle of the strata in the region.

Underground water

Often the landscape that forms from a particular kind of rock depends on the way water moves through it or over it. Sandstone, for instance, is a permeable rock. This means that it allows water to seep through it easily. This is not the same as being porous, though the terms are often confused. Porosity is the capacity of a rock to hold water in spaces in the rock – in other words, how full of holes it is. A rock like slate is barely 1 per cent porous; gravel is over 30 per cent porous. A rock that is porous is likely to be permeable – though not always – but a rock that is permeable is not necessarily porous.

Since sandstone is permeable, most water in sandstone landscapes soaks into the ground rather than flowing overland. This often makes sandstone landscapes quite angular because they are not rounded by water flowing over the land, except in times of flood. In wet regions, there may be too much water for the rock to soak up, so water erosion can do its work anyway.

Clays and shales, on the other hand, tend to be impermeable. They are often even more porous than gravel, but the pores are so small that water gets trapped. This means that the rocks easily become waterlogged and water remains on the surface to wear the rocks away, and round off contours.

Chalk and limestone

Like clay, chalk is not at all porous, and often contains clays which make it even harder for water to seep through. The same is true of limestone. Yet both these rocks are permeable. In particular, limestone usually has joints (cracks) so big that water can filter into the formation even if it can't penetrate the rock directly. It is a particular feature of limestone that the infiltrating water, called groundwater, corrodes away the rock to open up potholes and caverns.

Chalk has far fewer joints and is much less permeable than limestone and is not so susceptible to corrosion. So it rarely develops anything more dramatic than the occasional cavern. All the same, the chalk downlands of southern England do have their own special features including dry valleys called bournes. Bournes look like river valleys but contain no river. They may have formed in times when the climate

was wetter. Large hollows called combes may have been formed either by springs in these wetter times, or by frozen surface debris melting after the ice ages.

Granite

Like limestone, granite is tough enough to form mountains. In cold regions, granite intrusions are often left standing proud as dramatic monoliths when exposed on the surface. Yet in warmer regions, granite's feldspar content makes it prone to chemical weathering. Feldspar corrodes quickly in warm water. As with limestone, joints in granite allow water to penetrate deep into the rock, so tropical granite is corroded underground. When the weathered rock is stripped away, it leaves outcrops called kopjes. Similar features called tors form in cooler regions, such as in England's Dartmoor. It's thought that these are the result of climate change.

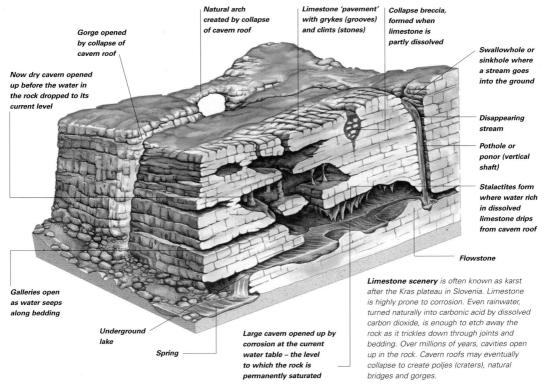

Gorge opened by collapse of cavern roof

Now dry cavern opened up before the water in the rock dropped to its current level

Natural arch created by collapse of cavern roof

Limestone 'pavement' with grykes (grooves) and clints (stones)

Collapse breccia, formed when limestone is partly dissolved

Swallowhole or sinkhole where a stream goes into the ground

Disappearing stream

Pothole or ponor (vertical shaft)

Stalactites form where water rich in dissolved limestone drips from cavern roof

Flowstone

Galleries open as water seeps along bedding

Underground lake

Spring

Large cavern opened up by corrosion at the current water table – the level to which the rock is permanently saturated

Limestone scenery is often known as karst after the Kras plateau in Slovenia. Limestone is highly prone to corrosion. Even rainwater, turned naturally into carbonic acid by dissolved carbon dioxide, is enough to etch away the rock as it trickles down through joints and bedding. Over millions of years, cavities open up in the rock. Cavern roofs may eventually collapse to create poljes (craters), natural bridges and gorges.

THE AGES OF THE EARTH

Written in the rocks of the crust is the entire history of the Earth's surface. The record is blurred in some places and lost in others. Nonetheless, by studying rocks in detail geologists have been able to piece together the remarkable story of how the Earth has changed through the ages.

Barely two centuries ago, most people thought the Earth was just a few thousand years old and little changed since it was created. But in the 19th century, geologists began to realize its age is immense. It is now thought to be nearly 4,600 million years old, and has undergone huge changes during its life. Its remarkable history is recorded in the rocks, if you know how to read it.

Pioneers of deep time

The great pioneers in the study of the Earth's distant past were James Hutton (1726–97) and William Smith (1769–1839). It was Scottish geologist Hutton who first suggested that the Earth is very, very old, and that the landscape has been shaped gradually by countless cycles of erosion and uplift.

Soon after, English surveyor William Smith was surveying routes for canals, and noticed that each layer of sedimentary rock contains its own range of fossils. He realized that all layers containing the same range of fossils must be of the same age. What's more, the 'principle of superposition' made it

possible to work out which layers are old and which young. In the 1600s, Danish geologist priest Nicolas Steno had realized that all sedimentary rocks were originally laid in flat beds, even if they've been tilted and broken since. Steno also realized beds were laid one on top of the other, so the oldest beds are always at the bottom and the

Above: Earth's rocks have been traced back over 3,800 million years. Rocks of this age are found exposed in western Greenland, like Sondre Strom fjord on the south-west coast.

youngest at the top. This is the principle of superposition.

Using this principle in conjunction with fossils, 'biostratigraphers' have built a detailed history of the Earth

Geologic time

This illustration gives a simplified picture of the major periods of Earth's history since the Cambrian Period 545 million years ago (mya) – the time when complex life forms first flourished.

4560–545 mya Precambrian The oceans formed and the first single-celled life forms appeared including algae which gave oxygen to the air. Later on, multi-celled animals like sponges and jellyfish appeared.

Blue-green algae (early life)

Trilobite (segmented creature)

Orthonybyoceras (marine animals)

Cooksonia (land plant)

545–495 mya Cambrian Period An explosion of life in the sea including small invertebrates and the first animals with hard shells easily preserved as fossils.

495–443 mya Ordovician Period Melting of the polar ice caps flooded much of the land. Crustaceans (like crabs), early marine animals and coral reefs appear.

Ichthyostega (amphibian)

Giant tree fern (first forest)

354–290 mya Carboniferous Period The sea level was high and large areas were covered in tree-filled swamps. Lime-stones laid down. Amphibians and insects spread, and possibly the first reptiles.

443–417 mya Silurian Period The Caledonian Orogeny threw up mountains in what became N America and NW Europe. Fish with jaws and river fish appeared, and the first land plants.

417–354 mya Devonian Period Continents and mountains grew. Old Red Sandstones laid down. Forests of club mosses and tree ferns spread. Vertebrates dominant. The first sharks. Animals began to live on land.

PALAEOZOIC

stretching back half a billion years. If rock sediments remained undisturbed forever, it would be possible in theory to slice through them to reveal most of the sequence of Earth's history. If you could take a column through the sequence, you could read Earth's history like a book.

The geologic time scale

Although such a column exists nowhere on Earth, a detailed geologic time scale based on it is now widely used. This timescale is constantly updated as geologists make new discoveries. It is detailed only back to the start of the Cambrian Period, 545 million years ago. Only since this time have shelly and bony life forms been common enough to leave a good fossil record. We once knew very little about the four billion years of Earth history before that, known as Precambrian time. In recent years, discoveries have begun to fill in the picture.

Just as day is divided into hours, minutes and seconds, so geological time is divided into units. The longest are Eons, lasting at least half a billion years. Eons are divided into Eras, Eras into Periods, Periods into Epochs, Epochs into Ages and Ages into Chrons. Each unit in the timescale is given a name,

usually derived from the area where rocks of the period were first studied. The Devonian Period, for instance, was named after Devon in England where rocks of the age were first studied.

Biostratigraphy now provides a detailed system for matching rock sequences around the world. Yet it has its limitations. It can show one rock is older than another, but not exactly how old. In other words, it gives a relative, not absolute, date. By working out how fast sediments might have been deposited and other clues, geologists established rough dates for the geologic timescale. But it is only with the development of radioactive dating in the last 50 years we can be confident dates are fairly accurate.

Radiometric dating

Atoms of elements can occur in alternative varieties or isotopes, each with a different number of particles in its nucleus. The number of particles is given in the isotope's name, such as uranium-235. Radioactive or radiometric dating uses the way certain isotopes 'decay' naturally through time – that is, break down to become isotopes of different elements. This

decay begins from the moment a rock is formed. It happens at such a steady rate that it is possible to work out how long it has been going on by counting the relevant 'daughter' isotopes in the rock compared to those of the original 'parent' isotope. The steady rate is known as the half-life, the time it takes for half the parent isotopes to break down. The widely used rubidium-87 isotope (which breaks down to strontium-87) has a half-life of 47.5 billion years. Rubidium is a rare element, but is often found with potassium in minerals such as feldspars and micas, and rubidium-strontium dating is used to date granites and gneisses.

Different isotopes are used to date rocks of different ages. Potassium-40, which decays to argon-40, is used to date rocks under a million years. Uranium-235 is used for the oldest rocks. Traces of uranium-235 decay products are found in zircon, one of the few minerals that survives for billions of years unchanged. So geologists studying ancient rocks look for zircon. Zircon found in the Jack Hills of Western Australia dates back 4300 million years, almost to the Earth's birth.

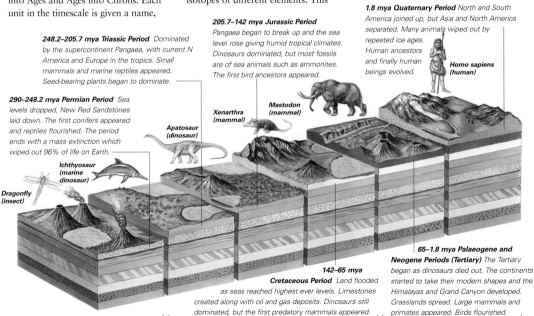

1.8 mya **Quaternary Period** *North and South America joined up, but Asia and North America separated. Many animals wiped out by repeated ice ages. Human ancestors and finally human beings evolved.*

Homo sapiens (human)

205.7–142 mya **Jurassic Period** *Pangaea began to break up and the sea level rose giving humid tropical climates. Dinosaurs dominated, but most fossils are of sea animals such as ammonites. The first bird ancestors appeared.*

248.2–205.7 mya **Triassic Period** *Dominated by the supercontinent Pangaea, with current N America and Europe in the tropics. Small mammals and marine reptiles appeared. Seed-bearing plants began to dominate.*

290–248.2 mya **Permian Period** *Sea levels dropped, New Red Sandstones laid down. The first conifers appeared and reptiles flourished. The period ends with a mass extinction which wiped out 96% of life on Earth.*

Mastodon (mammal)

Xenarthra (mammal)

Apatosaur (dinosaur)

Ichthyosaur (marine dinosaur)

Dragonfly (insect)

142–65 mya **Cretaceous Period** *Land flooded as seas reached highest ever levels. Limestones created along with oil and gas deposits. Dinosaurs still dominated, but the first predatory mammals appeared.*

65–1.8 mya **Palaeogene and Neogene Periods (Tertiary)** *The Tertiary began as dinosaurs died out. The continents started to take their modern shapes and the Himalayas and Grand Canyon developed. Grasslands spread. Large mammals and primates appeared. Birds flourished.*

MESOZOIC

CENOZOIC

ROCKS AND FOSSILS

The fossils that are found in most sedimentary rocks are one of the geologist's most valuable clues to Earth's history. Fossils not only help a geologist tell the age of a rock, but work out the conditions under which the rock formed – and track the occurrence of different strata right across the world.

Fossils are time capsules buried within almost every sedimentary rock. By identifying fossils and comparing them with creatures alive today, geologists can trace the way plants and animals have changed through time and use this knowledge to learn about the rocks they are found in.

The fossil record from the time before the Cambrian Period, beginning 545 million years ago, is sparse. Since then, however, millions upon millions of species have come and gone – many, many times more than are alive on Earth today. Probably only a tiny minority of these species have been preserved as fossils. But there are enough to provide a rich source of information.

'Fossil correlation' is one of the central techniques for geologists studying the history of rocks. Fossil correlation means tracing the occurrence of particular rock

Below: These arietitid ammonite fossils, in large pebbles washed ashore at Lyme Regis in Dorset, England, give very precise rock dating. They are an estimated 180 million years old.

formations over wide areas by looking for repeat occurrences of the same range of fossils. Sedimentary rock formations can be widely separated, but if they contain the same range of fossils, a geologist can be confident that they formed at the same time.

Moreover, since species change with time, many only appear in a particular part of the sequence of rock layers. So geologists look for the level where particular species first appear in the rock sequence, and for the point where they finally disappear. That way they can tell the relative ages of rock strata from the fossils they contain. Rock strata containing fossils of species that became extinct 300 million years ago are clearly older than strata containing fossils of species that first appeared about 250 million years ago.

Index fossils

Palaeontologists get excited if they find a rare dinosaur skeleton, but it is different for geologists. A fossil is little use for dating if it occurs in only a few rocks, if it is hard to identify, or if it

Above: If a sediment has only been buried at a shallow depth and little disturbed since this time, fossils like this bivalve may be preserved so intact that they look as if they have just fallen into the sand.

changed so little over time that it appears the same in rocks of all ages. So geologists actually look for certain very commonplace kinds of fossil, which they call 'index' fossils. If they spot one of these index fossils in a rock layer, they can often pinpoint its age instantly.

For a fossil to be used as an index, it must be widely distributed and easy to identify. It must also be small and have evolved rapidly, showing clear changes through time. All index fossils are small sea creatures – shellfish in

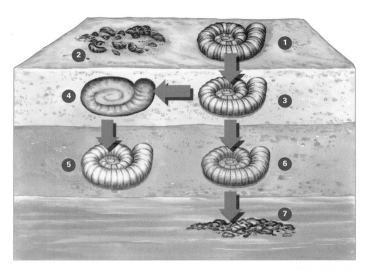

The fossilization process

Most fossils are shells or isolated bones. Complete skeletons are rare, and the soft body parts are almost never preserved. The vast majority of fossils are shellfish that lived in shallow seas. Fossils of land creatures are rare because they rotted away before they were preserved. When a creature such as a shellfish dies and falls to the sea floor, its soft body parts quickly rot away (1). However, its hard shell may be either broken (2) or buried intact (3). Over millions of years, the shell, made mostly of the mineral aragonite, may be dissolved away by water seeping through the sediments. Sometimes this leaves an empty cast or mould (4). The dissolved aragonite often recrystallizes in the cavity instead, creating a perfect replica of the shell in silicate or iron sulphide minerals (5). Occasionally, the shell is preserved almost intact (6). Nearly all fossils are destroyed if buried too deep or the rock undergoes metamorphism (7).

particular – whose remains are easily preserved in the sea-bed sediments that form most sedimentary rocks.

Fossil zones

Using these index fossils, geologists try to divide local sequences of rock into 'zones', each containing a particular range of index fossils. The acme zone is where the index fossil is especially abundant – either because the species thrived at the time the sediments were laid down, or because they happened to be especially well preserved. The full range of strata through which the fossil is found is called the range zone. When two or more species of index fossils overlap in the sequence, they are said to form a concurrent range zone. Where one of the index fossils is not available, geologists rely on fossil assemblages. These are groups of fossils known to have lived at the same time, for example dinosaurs, redwoods and dragonflies.

Zonal schemes

Because most creatures can only survive in a particular range of environments, most appear only in particular kinds of rock. So each kind or 'facies' must have its own zonal scheme. This takes into account the range of creatures living at that particular time in each rock-forming environment. In Devonian rocks in Europe, for instance, there are three main facies, each with its own zonal scheme. Old Red Sandstone, which formed in lakes and estuaries, is zoned or indexed by fossil fish. Rhenish rock which formed on warm, shallow sandy sea-beds is zoned by brachiopods and corals. Hercynian rocks, which formed on deep muddy sea-beds, are zoned using ammonites.

Index fossils

The illustration shows some of the most widely used index fossils around the world. Below are listed the key marine fossils used in North America, two for each period. ID pictures can be found in a good fossil guide. Note that the Carboniferous Period is divided into the Mississippian and Pennsylvanian in North America.

Quaternary	Pecten gibbus	Neptunea tabulata
Tertiary	Calyptraphorus velatus	Venericardia planicosta
Cretaceous	Scaphites hippocrepis	Inoceramus labiatus
Jurassic	Perisphinctes tiziani	Nerinea trinodosa
Triassic	Trophetes subbuliates	Monotis subcircularis
Permian	Leptodus americanus	Parafusulina bosei
Pennsylvanian	Dictyoclostus americanus	Lophophyllidium proliferum
Mississippian	Cactocrinus multibrachiatus	Prolecanites gurleyi
Devonian	Mucrospirifer mucronatus	Palmatolepus unicornia
Silurian	Cystiphyllum niagarense	Hexameroceras hertzeri
Ordovician	Bathyurus extans	Tetragraptus fructicosus
Cambrian	Paradoxides pinus	Billingsella corrugata

Ammonites were squidlike shellfish that make good index fossils for the Jurassic and Cretaceous Periods.

Trilobites were one of the first creatures with a body split into segments. They are index fossils for the Cambrian.

Echinoderms are very common fossils but have changed so little through the ages that they are of little use as indexes.

Corals are very common fossils but have changed so little through the ages that they are of little use as index fossils.

Graptolites were sea creatures that make good index fossils for the Ordovician and Silurian Periods.

HOW MINERALS FORM

Minerals are the natural substances the world's rocks are made of. All of them are solid crystals with a particular chemical composition – some so tiny they can only be seen under a powerful microscope, others as big as tree trunks. Each kind forms under particular conditions in particular places.

There are 4,000 to 5,000 different minerals in the Earth's crust. Yet only 30 or so are very widespread. Most of the rest are present in rocks only in minute traces, and are only easy to see when they become concentrated in certain places by geological processes. It is concentrations like these that give us the ores from which many metals are extracted.

Furthermore, large crystals of minerals like those that illustrate this book, even common minerals, are so rare that a mineral hunter feels understandably excited to find one. Big, spectacular crystals need both time and space to grow – and a steady supply of exactly the right ingredients. Such a perfect combination is extraordinarily rare.

Mineral crystals form in four main ways. Some form as hot, molten magma cools and crystallizes. Some form from chemicals dissolved in watery liquids. Some form as existing minerals are altered chemically, and some form as existing minerals are squeezed or heated as rocks are subjected to metamorphism.

Above: This woman is inside the world's biggest geode – most are no bigger than a fist. Geodes are cavities that probably form from gas bubbles in lava (or limestone). The bubbles later fill with hydrothermal fluids in which large crystals such as amethyst grow.

Below: The best crystals need space to grow, so they are often found in cavities, such as geodes. Geodes can look like dull round stones on the outside, but a tap with a hammer gives a tell-tale hollow sound. When cracked open, they reveal a glittering interior.

Minerals from magma

As magmas cool, groups of atoms begin to come together in the chaotic mix and form crystals. The crystals grow as more atoms attach themselves to the initial structure – just as icicles grow as more water freezes on to them. Minerals with the highest melting points form first, and as they crystallize out the composition of the remaining melt changes (see How Igneous Rocks Form, this section).

Chemicals that slot easily into crystal structures are removed from the melt first, and it is bigger, more unusual atoms that are left behind. It is these 'late-stage' magmas, the last portion of the melt to crystallize, that give the most varied and interesting minerals.

Just what these minerals are depends on the original ingredients in the magma, and the way it cools. Large crystals tend to form in magmas that have cooled slowly. The biggest and most interesting often form in what are called pegmatites, which form from the fraction of melt left over after the rest has crystallized. Pegmatites typically collect in cracks in an intrusion or ooze into joints in the country rock, forming sheets of rock called dykes. The residual fluids in these late-stage magmas are rich in exotic elements such as fluorine, boron, lithium, beryllium, niobium and tantalum. These can combine to form giant crystals of tourmaline, topaz, beryl, and other rarer minerals. When the fluids are rich in boron and lithium, tourmaline is formed. When fluids are rich in fluorine, topaz is formed. When fluids are beryllium-rich, beryl forms.

Minerals from water

Water can only hold so much dissolved chemicals. When the water becomes 'saturated' (fully loaded), the chemicals precipitate out – they come out of the water as solids. Typically this happens when water evaporates, or cools down.

When sodium, chlorine, borax and calcium are dissolved from rocks, they may be carried by rivers to inland seas and lakes, which then evaporate, leaving mineral deposits of minerals such as salt, gypsum and borax.

Many other minerals form from the cooling of hydrothermal solutions – hot water rich in dissolved chemicals. Sometimes the water is rainwater that seeps down through the ground (meteoric water) and is then heated by proximity to either the mantle or a hot igneous intrusion.

Hydrothermal solutions also come from late-stage magmas, and so are rich in unusual chemicals. Such solutions ooze up through cracks in the intrusion and cool to form thin, branching veins.

Alteration minerals

Although some minerals such as diamond or gold seem to last forever, most have a limited life span. As soon as they are formed, they begin to react with their environment – some very slowly, some quite quickly. As they react, they form different minerals.

Metal minerals are often oxidized when exposed to the air, or oxygen-rich water. Iron minerals rust, like iron nails, turning to red and brown iron oxide. When water containing dissolved oxygen seeps down through the ground into rocks and veins containing metals, it creates an oxidation zone in the upper layers as the metals are altered. Cuprite, goethite, anglesite, chalcanthite, azurite and many other minerals form this way. Some sulphide minerals are oxidized to sulphates that dissolve in water. These sulphates may be washed down through the rock to be deposited lower down as different minerals which become often valuable ores such as chalcocite.

Minerals remade

Many minerals become unstable when exposed to heat and pressure and respond by altering their chemistry to become different minerals. This is known as recrystallization and is typically linked to metamorphism. When a rock is remade by metamorphism, its mineral ingredients are recrystallized.

While it takes hot magma or tectonic movements to metamorphose rocks, simple burial is often enough to alter minerals, since heat and pressure rise as depth increases. Minerals can also be recrystallized by contact with hydrothermal fluids.

In the simplest recrystallizations, the resulting minerals depend merely on the combination of heat and pressure, and the minerals in the original rock. However, new ingredients may seep in to alter the picture. Where magma intrudes into a limestone, for instance, the magma 'cooks' the limestone but also introduces new chemicals to create a complex mix. The limestone supplies calcium, magnesium and carbon dioxide, while the magma brings silicon, aluminium, iron, sodium, potassium and various other ingredients. The result is a 'skarn' rich in a huge variety of interesting silicate minerals.

Above: Topaz is one of many rare minerals that form when the last rare-mineral-rich residue of a magma finally crystallizes, particularly in dykes called granitic pegmatites. Topaz forms from residues rich in fluorine. Crystals formed in voids in the pegmatite, called miarolitic cavities, can occasionally grow very large.

Above: Fluorite is one of many minerals that form as hot hydrothermal solutions cool and precipitate some of their dissolved chemicals. Natural pipes carrying hydrothermal veins eventually completely crystallize to form veins.

Above: Cuprite is one of many minerals that form by oxidation reactions, caused by exposure to air, or oxygen-rich water. Cuprite forms a bright green crust on oxidized copper minerals.

Left: Rubies are among a number of rare and precious gem minerals that form when certain oxides are crystallized by the heat and pressure of metamorphism.

GEOLOGICAL MAPS

A good geological map is probably the geologist's single most valuable tool. It displays what rocks appear where in the landscape and the major geological features. This not only helps the geologist identify rocks on the ground but provides a good indication of where to find particular minerals.

For the experienced geologist, a geological map does not simply show where rocks appear. With skilful interpretation, it is possible to work out the three-dimensional structure of rock formations, the way they relate to each other, and even some of their history and the way that the landscape has developed.

Most conventional geological maps are known as 'solid' maps, because they show the solid rocks under the surface. They help geologists to find the rock structures most likely to yield mineral ores and oil and gas deposits.

Geological maps can also show loose surface deposits, such as the sediment deposited by rivers in flood

Mapping the landscape
The same area can be mapped by geologists in four different ways. A satellite image (left) shows the landscape very clearly, allowing the geologist to interpret ground features relating to the underlying geology before surveying in the field. A step on from the satellite image is the digital terrain model (below), which combines satellite and ground survey data to build a 3D computer model. This gives an even clearer picture of the landscape and clues to the underlying geology.

or by glaciers. Maps like these are called 'drift' maps and are valuable for the construction industry since they reveal just how solid the ground on which they plan to build is likely to be. They are invaluable in the initial plotting of the course of a railway tunnel, for instance, before detailed survey work on the ground begins.

Colours and symbols
Geological maps usually use areas of different colours to represent the different kinds of rock or drift. Igneous rocks, for instance, are usually shown in various shades of purple and magenta, depending on whether they are intrusive or extrusive. Metamorphic rocks are typically shown in shades of pink and grey-green. Sedimentary rocks are usually shown, appropriately, in sandy colours – shades of brown and yellow, plus green – except for limestones, which are typically blue-grey.

Besides colours, maps often use types of shading or pattern called ornament. Each rock type is also given a set of letters to symbolize it on the map. This usually begins with a capital letter to show the age of the rock – for example J for Jurassic, K for Cretaceous, T for Tertiary and Q for Quaternary. The smaller letters indicate either the name of the particular formation or rock type.

Geology underground
Geological maps only actually show surface geology. Even solid maps only show outcrops – that is, where a solid rock of a particular kind reaches the surface, even if it is actually hidden beneath sheets of deposits. Outcrops that are not covered with deposits are called exposures. All the same, many maps have cross-sections showing a vertical slice through the rock formations, revealing how they are

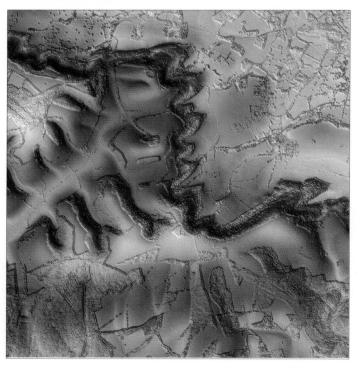

arranged beneath the surface. Often these incorporate results from boreholes and seismic surveys. Yet an experienced geologist can construct a cross-section directly from the map.

Where there are a series of roughly parallel bands of sedimentary rocks, for instance, it is highly likely that they are simply where gently dipping (tilted) sedimentary strata reach the surface. Contours on the map reveal the shape of the land surface. If these reveal an escarpment with a gentle dip slope (see Rock Landscapes in this section), the angle of the dip (tilt) is easy to guess.

Geology in 3D

Cross-sections only show the structure of rocks in the landscape in two dimensions – as a thin slice – but

A third map (below) shows the rock underlying the same terrain, with different rocks indicated by coloured bands. A professional geologist can use this map to construct a cross-section (right) of the landscape from, say, A to B, plotting the height given by contours, then interpreting how the rock beds lie from their surface configuration.

geologists often want to know the whole structure of the area in three dimensions. In the past, they often constructed models called fence diagrams. These were made by constructing a series of cross-sections at right angles, interlocking like the fences of a square field. Sometimes these were created with card, and sometimes they were simply drawn.

Computer modelling

With the spread of computers, however, geologists are able to construct complete 3D models of the ground. These can be manipulated and

projected to reveal the subsurface structure in any way the geologist wants. Many geologists hunting for minerals now use these computer models instead of conventional maps wherever they exist.

Until recently, it was solid geology, not drift geology that received the attention from computer modellers. Now, however, organizations such as the British Geological Survey (BGS) use GSID (Geological Surveying and Investigation in 3D) software to create 3D models that show not only outcrops in their entirety but the surface deposits as well.

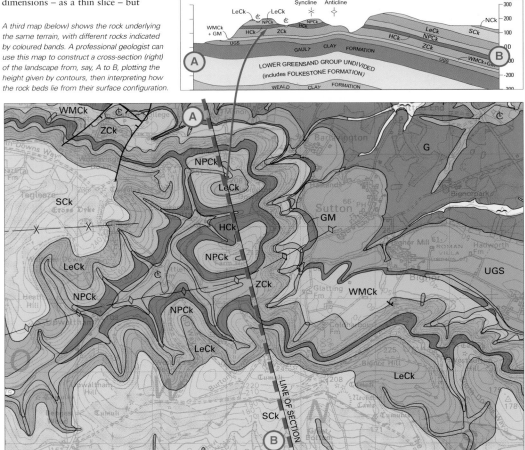

READING THE LANDSCAPE

Every bit of rock has a story to tell. Look at a craggy cliff face or pick up a stone on the shore, and if you know what to look for, you can read in it a great deal of its past, from scratches revealing episodes of glaciation to rock boundaries marking cataclysmic earthquakes.

In some ways, the geologist is like a detective. The idea is to look for clues, study the evidence and work out just what has happened. Sometimes the clues are so small they are visible only under a powerful microscope. Sometimes they are so big they are visible only from space. Fortunately most can be identified with no more specialist equipment than basic common sense.

If you find a stone, see what you can work out about its history. At first it might seem baffling and dull, but with a little thought you can often piece together some of its long story. Obviously it helps to identify the stone, and the Directory of Rocks later in the book should help you here. Yet even without a strong positive identification, you can often work out quite a bit about it.

First of all, you can probably guess whether the stone arrived naturally where you found it – or whether there was a human hand involved. If you see other similar stones nearby in a natural setting, the chances are it got there naturally. Exotic stones that look very different can arrive in their resting

The Great Unconformity
One of the world's most famous unconformities is North America's Great Unconformity. This dramatic gap in the geological sequence stretches all the way from Arizona to Alberta in Canada. Perhaps the best place to see this is in the Grand Canyon National Park (the Canyon's west rim is shown above), where the Colorado River has cut down through the strata over five million years to reveal the Unconformity at the canyon's foot. Here young Tapeats sandstone sits so directly on top of ancient, two-billion-year-old Vishnu schist that it is possible to touch them both with the span of a hand. The schists began life some two billion years ago as sediments. These sediments were metamorphosed to schist

about 1.7 billion years ago as they were penetrated by a magma intrusion and squeezed by tectonic movement. Slowly, over more than a billion years, this schist was worn flat, then sank beneath the sea. About 500 million years ago, the sediments that formed the Tapeats sandstone began to pile up on the sea floor. The sea floor began to subside under the weight of sediments, and sedimentation continued over hundreds of millions of years with little or no tilting, Eventually some 1,220 m/4,000 ft of level sedimentary beds were laid down. Since then the whole area has been uplifted, and the path of the Colorado river has exposed the whole sequence to reveal the Tapeats sandstone and the Great Unconformity at its base.

Below: Cliffs and chasms, such as this feature of the Blue Mountains in NSW, Australia, are one of the best places to see rock strata.

place naturally. Glaciers, for instance, carry entire giant boulders, called erratics, into areas of different rock. Flash floods and avalanches can carry quite large stones. Yet exotics like these are rarities. Most loose stones in a natural setting are not only similar to each other; they are often similar to the rocks nearby. If they resemble the solid rock around, you can be sure the stones were eroded from it.

Often, you can see where the stones came from immediately. Reasonably large stones are rarely carried far from their origin. Often, stones simply fall down a mountain slope to gather in a scree at the foot. Stones and pebbles can fall from a sea cliff on to the beach

below. There is often a marked difference between stones in a scree and stones on a beach. Scree stones are almost invariably sharp-edged and chunky, reflecting the way they were shattered from the mountain above by frost, often within the past few months. Beach stones, however, are often rounded pebbles. They are rounded because they have been rolled over and over in water laden with abrasive sand again and again. To get so rounded, they must have been in the water many thousands of years. Waves are powerful enough to carry stones some distance, so the pebbles on a beach have not necessarily come from the cliffs above.

How an angular unconformity forms

Sediments are usually laid down in a continual sequence but sometimes a break or unconformity develops. Shown below are the stages in the creation of an angular unconformity. It begins with the laying down of a sequence of sediments (1) which are then uplifted and folded into mountains (2). The mountains are worn down to a plain over a long period (3). The sea level rises, submerging the plain (4). New layers are laid down horizontally in the sea, over the contorted layers of the original sediments (5).

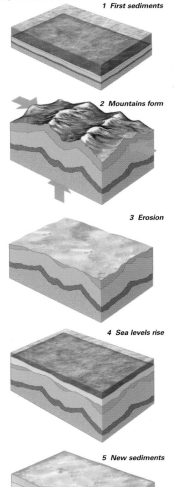

1 First sediments

2 Mountains form

3 Erosion

4 Sea levels rise

5 New sediments

Using rock layers

On a larger scale, you can tell a great deal about the history of rocks, especially sedimentary rocks, by stratigraphy – the study of rock layers and their contents. Earlier pages have shown how if rocks contain the same assemblage of key fossils they must be the same age. The law of superposition also shows how the uppermost layers are likely to have formed later in Earth's history. Another valuable rule of thumb is that the more contorted strata is, the older it is likely to be. Similarly, the law of cross-cutting relationships shows that whenever a geological feature like an igneous intrusion cuts across a sequence of rock layers, it is certain to be more recent than the rock layers.

Breaks in the sequence

One feature geologists look for in particular is an unconformity. On the whole, sediments are laid down one on top of the other in sequence, but every now and then you find a gap or even a dramatic break in the sequence. A young rock may overlay an old rock, with no intervening 'middle-aged' rocks, or a sedimentary sequence may overlay metamorphic rock. This gap or break is the unconformity, and it can be very revealing.

One kind of unconformity, easily seen in the field, is an angular unconformity. In this case, younger sediments sit on top of older sediments that have been tilted and deformed and then eroded to form a flat plain.

There are other kinds too. If the older sequence remains both level and undeformed, then the only sign of the unconformity may be the great age difference between rock layers, indicated by a significant gap in the fossil record. This is known as a parallel unconformity.

A disconformity is formed where the eroded surface of the older sequence is not a flat plain but has hills and valleys. In this instance the break in the sequence is marked by a corresponding wavy line. A non-conformity occurs where the sedimentary rock sequence has been interrupted by either igneous or metamorphic rock.

Stratification

When a geologist sees a sequence of rocks exposed in a cliff, he can examine the layers and build up a picture of their history, using familiar clues. The nature of the rock gives a clue to the environment in which they formed. Fossils in the rock layers may give a clue to their relative age. So too can their position within the sequence, while unconformities bear witness to dramatic breaks in the sequence, and past geological events.

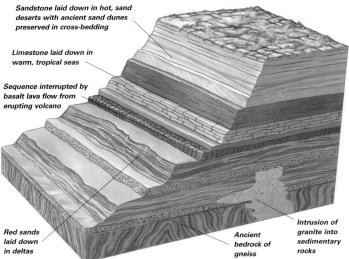

Sandstone laid down in hot, sand deserts with ancient sand dunes preserved in cross-bedding

Limestone laid down in warm, tropical seas

Sequence interrupted by basalt lava flow from erupting volcano

Red sands laid down in deltas

Ancient bedrock of gneiss

Intrusion of granite into sedimentary rocks

MINERAL ORES

Our modern society could not function without metals extracted from rocks, yet even aluminium and iron, the most abundant metals, make up just a few per cent of rocks by weight. Fortunately, these and many other useful metals are concentrated by geologic processes in a few places as 'ore' minerals.

The first metals people used were native metals – copper, silver and gold – which occur as chunks of metal in the ground and could be fashioned into jewellery and knives. Then some unsung genius realized that there are minerals that while not actually metallic contain metals that can be extracted by heating until the metal melts out. This smelting process has provided us with metals ever since.

Minerals which contain enough metal for it to be easily extracted are called ore minerals. Galena contains a very high percentage of lead, so is a major lead ore. Iron comes mainly from hematite and magnetite. Pyrite is also rich in iron, but because it is bound so tightly to sulphur in the mineral it cannot be extracted, so pyrite is rarely an ore of iron.

To form an ore, the metal must be sufficiently concentrated in the mineral, and the mineral sufficiently concentrated in the ground for it to be worth extracting. Fortunately, geological processes ensure that such concentrations, known as ore deposits, do occur in certain places.

Below: Valuable minerals such as gold may sometimes be concentrated in iron-bearing ore deposits called gossans. They get their name from the Cornish for 'blood' because of their rusty red colour from the oxidized iron.

Above: Most ores are extracted from near-surface deposits in vast opencast mines like this copper mine in Bisbee, Arizona. Huge quantities of rock have to be dug out, much of which is later discarded.

Hot deposits

Many important ore deposits are linked to magma chambers, where melted rock collects. As the magma cools and begins to solidify, heavier minerals begin to sink to the bottom of the chamber. So when the magma finally freezes solid, heavy ore minerals may be concentrated at the base, especially sulphides ores, creating what are called magmatic sulphide ore deposits. Deposits like these tend to form only in relatively runny magmas such as komatiite basalt and gabbro. Famous sulphide deposits of this kind include South Africa's Bushveld Complex, Noril'sk in Russia, Jinchuan in China and Sudbury in Canada. Although their origins are not entirely understood, massive sulphide deposits can also occur in rock beds metamorphosed in the granulite facies (see Schists: Metamorphic facies, World Directory of Rocks). Australia's Broken Hill has superb lead, silver and zinc ores like this.

Rich ore deposits are also created by hydrothermal (hot-water) solutions circulating through magma or the rocks surrounding an intrusion. These fluids are typically rich in dissolved metals, which they deposit in fractures and pores, creating hydrothermal deposits.

If a hydrothermal deposit is dispersed throughout the intrusion, it forms a 'disseminated' deposit. If it follows cracks in the rock, filling them in with deposited minerals, it forms a hydrothermal vein deposit. Vein minerals include sulphide, oxide and silicate ores, as well as native metals such as gold, silver and platinum. Gold often appears as flakes in white quartz veins. Copper ores are often concentrated in porphyritic intrusions, forming porphyry copper deposits.

Some of the most valuable deposits are created when hydrothermal solutions percolate into limestone. When solutions seep into limestones and marbles around an intrusion, they create skarns, such as the tungsten skarns of Sangdong, Korea, King Island, Tasmania and Pine Creek in California.

The same process creates Mississippi Valley Type (MVT) deposits, which occur around the edge of sedimentary basins, deep down at the base of limestone beds. MVTs formed as hot solutions seeping through underlying rocks reacted with the limestone. North America's Tristate zinc district is the best known MVT, but there are also well known MVTs in Cumbria, England and Trepca, Serbia.

The hot waters flushed out of submarine volcanoes along the mid-ocean ridge are often rich in dissolved metals and sulphur. When the hot water meets the cold ocean, these dissolved chemicals often combine to form deposits of metal sulphides, which may be recovered in places where tectonic movements have lifted up ancient sea floor to much more accessible places.

Cool deposits

It is not just hot water that concentrates minerals. Even cool groundwater can dissolve metal ores as it seeps through rocks. The metal ores may have been widely scattered through the rock, but when the water re-deposits them, it may deposit them together in a concentrated deposit – a

Right: Lode gold is often found associated with quartz in veins like this one in siltstone in the Tanami desert of Australia's Northern Territory. This region is known for its gold and has recently become the target of seismic profiling, designed to reveal its geological structures and reveal likely sites for gold from the pattern of powerful vibrations sent through the ground by machines.

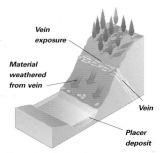

Placer deposits As veins are exposed at the surface, valuable minerals such as gold may be tumbled or washed down slope to collect in stream beds, forming placer deposits. Prospectors may sometimes recover these placer golds by panning the stream gravel.

process called secondary enrichment. This often happens when water seeping down through the ground reaches the water table, the level to which the ground is saturated.

In the opposite way, rainwater sinking into the ground may leach (dissolve) away particular chemicals, leaving behind concentrations of metals such as iron and aluminium. In the tropics where rainfall is heavy, the concentrations in these 'residual deposits' can become so intense that the soil itself becomes an ore, like bauxite aluminium ore.

Water can even help concentrate ore deposits without dissolving anything. After weathering has broken rocks up, rivers and streams carry away mineral grains. Heavier, more durable grains are dropped first by the water, and can accumulate in riverbed deposits. Gold, tin, diamonds and emeralds are among the minerals found in 'placer' deposits.

Some of the world's most important cold water ores formed in seawater. These include nodules of deep-sea manganese, deposited on ocean beds, and the remarkable Banded Iron Formations (BIFs) – rocks containing colourful layers of iron oxides and sedimentary rocks formed by bacteria billions of years ago, and now one of the world's major sources of iron.

Finding ores

In the old days, prospectors would scour the landscape hoping to stumble on 'shows' – exposures of ores on the surface, with little to guide them other than a knowledge of what geologic features they are often found in association with. Nowadays, geologists still read the landscape but have an array of sophisticated technology to help them, including aerial and satellite photographs. Kimberlite pipes that might yield diamonds often show as pale discs on the surface, for instance.

Once a potential site has been identified, the extent and richness of a deposit can be assessed by testing the ground's electrical conductivity and its magnetism. An instrument called a magnetometer may be used to locate deposits of magnetic ores like magnetite and ilmenite. Since ore minerals tend to be denser than average, measuring the local pull of gravity can also be revealing. Radioactive minerals and elements such as uranium and thorium might be detected with a Geiger counter. Since plants absorb traces of metals through their roots, it can even be worth analysing the plants in the area.

COLLECTING ROCKS AND MINERALS

You need very little special equipment to start building a rock and mineral collection – just sharp eyes for loose specimens when you're out walking. All the same, it helps to know where to look, and to acquire a few basic tools and storage systems for extracting good specimens and looking after them.

You can see rocks and minerals in many places. Office blocks often have polished granite faces. Houses might be built in sandstone and roofed with slate. Statues may be carved from marble. People wear precious gems made from mineral crystals. Many enthusiastic geologists get satisfaction from spotting such occurrences and identifying the rock.

Some enthusiasts build up a collection by looking for samples at rock shops and on the internet. But there is nothing to beat building up your own collection, by going out into the 'field', as geologists call it, to find your own samples.

Good, collectable specimens are not evenly spread around the Earth, but are concentrated in particular sites.

Some sites yield one or two kinds of mineral; others several hundred. The more you get to know local geology, the more likely you are to find good specimens. Good crystals are found in cavities and fissures, gemstones in pegmatites, gold in milky quartz veins and so on. Often minerals on the surface may be signposts to the real treasure beneath, like green copper.

What you need

Hammer
The key item in the geologist's toolbag is a good hammer. You can manage with an ordinary bricklayer's hammer, but it is really worth investing in a proper geological hammer. This typically has a square striking face and tapered tail (7). In the most common 'chisel' hammers, the tail is a flat edge at right angles to the handle, useful for levering samples out. In 'pick' hammers, the tail is a long, curved cutting edge, and is good for splitting rocks (6).

Chisel
There are three kinds of chisel. Cold chisels have a long, narrow handle and wedge, ideal for extracting crystals from cavities (8). Gad-point chisels are shorter and thicker, with a tapered point good for splitting, prising and wedging rocks apart. Broad-bladed chisels, bolsters and 'pitching tools' (10) are short and thick with a wide blade, good for splitting and trimming rock samples. If you only have one chisel, get a gad-point.

Safety equipment
Rocks can splinter when hit, so it is vital to wear good goggles when hammering. Goggles should have clear, hard plastic windows and flexible plastic sides (5). Avoid goggles without sides. Strong leather gloves are also useful protection and if you are visiting cliffs or quarries a safety helmet to protect the head is essential equipment (1).

Magnifying lens
A good magnifying glass or hand lens (3) not only helps you spot small crystals in the field but also helps you identify minerals and rocks you have found. A 3-times magnification lens is too weak to be of much help, but don't make the mistake of getting too powerful a lens. A 20-times gives too restricted a view to be useful in the field. The best compromise is a 5- or 10-times lens. Lenses are easy to lose, so tie your lens on a cord and attach it to your belt or collecting bag, or hang it round your neck.

Note-taking and recording
A notebook (2) and pen (9) are still the best way of noting down your finds. Sticky labels and markers (4) are good for labelling them on the spot. A small camera is useful for recording the site, and sometimes saves digging out a sample unnecessarily.

Way-finding
You need good local maps and perhaps geological maps to help you find your collecting site. A handheld Global Positioning System (GPS) is also useful for finding your way if you're straying far from the beaten track, and for recording precise locations and details of finds. A compass is a useful alternative – and enables you to test minerals for magnetism.

Collecting equipment
You need a strong bag for your samples. A small rucksack is ideal, leaving your hands free to pick up samples. Take bubble wrap and freezer bags to wrap samples.

Extra tools
In terms of additional portable items, a multi-purpose penknife and a small shovel are useful but not essential items. Shovels should have a shallow, pointed blade.

Hunting for rocks

Beaches and sea cliffs are good places to look for samples. The sea not only exposes the rock layers but pounds specimens free for you. But safety precautions are paramount. Always work at the cliff foot – never climb up – and wear a hard hat to protect your head from falling rocks. When hunting samples anywhere but a beach, always remember that most land belongs to someone and you may need permission to collect samples. Indeed, any samples found on someone else's land belong to the owner of the land. It's especially important to get permission if hunting in old quarries. Check also if the site is a nature reserve or protected in some way. Collection of samples may be forbidden. Even where you are allowed to collect samples, always be considerate. Don't ruin the site for others, clean out all samples or damage the rock.

Finding and cleaning specimens

There are two kinds of site to look for specimens: rock outcrops and deposits. Rock outcrops include cliffs, crags, quarries and cuttings. Deposits are where loose rock fragments have accumulated, including river beds, beaches, fields and even backyards. They include placer deposits where nuggets of gold and gems such as diamonds have come to rest in stream sand and gravel, and which continue to attract serious prospectors.

Specimens taken from the ground are often filthy and should be cleaned before you stow them away. Make sure you have identified the mineral first, however, as some minerals, like halite, are soluble in plain water! For soluble specimens, scrub with a toothbrush, or dab with pure alcohol. If the sample is very soft, just use a blower brush like photographers use to clean lenses.

Below: The geologist's rule is to hammer as little as possible. Hammering promotes erosion and leaves a scar in the rock. You should never use a hammer for knocking out samples – only for breaking specimens up. Always wear goggles to protect your eyes from splinters.

With most minerals, fortunately, you can brush off loose dirt with a soft toothbrush, then rinse in warm (not hot) water. For greasy marks and stains, add a drop of household detergent to the water. If the specimen is encrusted with mud and grit, don't try to chip it off. Leave it to soak overnight to soften. It is fine to attack hard specimens like quartz with a nailbrush, but delicate specimens like calcite are all too easily damaged.

You can use vinegar to dissolve away unwanted calcite and limey deposits on most insoluble minerals. Iron stains can be removed with oxalic acid. You can get this from chemists, but it is poisonous so should be handled with care. It dissolves some minerals so test it on a fragment first.

Once your specimen is clean, you need to put it away in a cool, dry, dark place to keep it at its best. Don't store different minerals touching each other. Rocks and minerals 'breathe' and absorb and emit gases over time, and may alter accordingly. Keep your collection away from windows, room heaters, humid places such as bathrooms, and car exhaust, and try to keep conditions as stable as possible.

Some minerals, such as native copper and silver, oxidize and tarnish, especially in polluted city air. Some minerals, like borax, dry out, so store them in air-tight containers. Halite absorbs water from the air and

gradually dissolves unless you keep it in an air-tight container with a little silica-gel to absorb any moisture.

Proper sample display drawers are expensive. Shallow drawers, wooden trays in cupboards, or glass and plastic boxes will do as alternatives. Identify every specimen with a number either on a sticky label or on a dab of white paint on an inconspicuous part of it. Enter this number, with the specimen profile, in your catalogue, which can be a computer file, cardfile or a notebook. You can group specimens by location or colour, but most geologists prefer to group by type: rocks into igneous, sedimentary and metamorphic types; minerals into chemical groups such as silicates and carbonates.

Cataloguing a collection

The minimum data for a catalogue is your own catalogue number, the mineral or rock name and for minerals a Dana number (see Classifying minerals in this section). But it is better to put as much data as you can. These are the details you should have:
1. Personal catalogue number
2. Dana number for minerals
3. Mineral or rock name
4. Chemical or mineral composition
5. Mineral or rock class
6. Exactly where you found it
7. Name of rock formation or kind of site where you found it
8. Date you found it
9. The collector's name (you)
10. Any other details, such as its history if you bought it, unusual characteristics of the specimen, and so on

CLASSIFYING ROCKS

Most geologists agree on the basic grouping of rocks into igneous, sedimentary and metamorphic rocks, but when it comes to classifying rock species within those three broad groups, there is a great deal of controversy and there is no definitive system.

Classifying rocks is not simply a matter of identifying rocks and sorting them. Each classification system depends on a theory of how rocks are made. As ideas of how rocks are made change with new discoveries, so too do rock classification systems.

Over the last half century, for instance, there have been over 50 different classification schemes suggested for sandstone alone – and sandstone is by no means the most contentious and complex of rocks. The debate over igneous rocks has been, if anything, more heated. So the classifications presented here are essentially just a snapshot.

Igneous rocks

Rocks of this type are perhaps the most complex of all rocks to classify. Yet there is a surprisingly simple and easy-to-use basic classification that works well in the field. This is one based on colour and texture.

The texture or average grain size of an igneous rock depends largely on how long the melt from which it formed took to cool. So rocks that developed deep in the earth such as granite are coarse-textured or phaneritic, while rocks that formed on the surface such as basalt are fine-grained or aphanitic. Rocks which are basically fine-grained but contain large

crystals (phenocrysts) are termed porphyries. Using texture, it is, on one level, easy to group igneous rocks according to their origin, into coarse-grained 'plutonic' rocks formed at great depth, mixed and porphyritic 'hypabyssal' rocks formed at shallow depths, and fine-grained 'volcanic' rocks formed from lava on the surface.

Colour provides the other means of basic classification. For reasons that are not entirely clear, minerals at the top of Bowen's Reaction Series (see How Igneous Rocks Form in this section) like pyroxenes and amphiboles tend to be dark in colour while those at the bottom like quartz and plagioclase

Classifying igneous rocks

Colour and texture provide a good basic classification of igneous rocks, but geologists often need a more detailed system based on composition since colour and texture are not always enough to distinguish rocks. The Streckheisen system groups rocks according to the proportion of four minerals they contain: quartz, alkali feldspar, plagioclase feldspar and feldspathoids (foids). The percentages of each can be plotted on the diamond shaped diagrams. The corners represent 100% of the appropriate mineral.

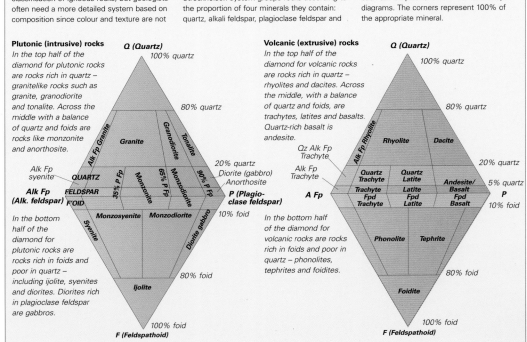

Plutonic (intrusive) rocks
In the top half of the diamond for plutonic rocks are rocks rich in quartz – granitelike rocks such as granite, granodiorite and tonalite. Across the middle with a balance of quartz and foids are rocks like monzonite and anorthosite.

In the bottom half of the diamond for plutonic rocks are rocks rich in foids and poor in quartz – including ijolite, syenites and diorites. Diorites rich in plagioclase feldspar are gabbros.

Volcanic (extrusive) rocks
In the top half of the diamond for volcanic rocks are rocks rich in quartz – rhyolites and dacites. Across the middle, with a balance of quartz and foids, are trachytes, latites and basalts. Quartz-rich basalt is andesite.

In the bottom half of the diamond for volcanic rocks are rocks rich in foids and poor in quartz – phonolites, tephrites and foidites.

Lutite (fine-grained):
shale

Arenite (medium-grained):
sandstone

Rudite (coarse-grained):
breccia

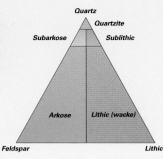

QFL composition triangle for
sandstone rocks

Classifying clastic sediments: texture
Above: The simplest way of classifying
clastic sediments is by texture into three
groups: lutites, arenites and rudites. Lutites
are mainly silt and clay particles; arenites
mainly sand; and rudites gravel, pebbles,
cobbles and boulders.

Classifying clastic sediments: composition
Right: They can also be divided according to
composition in terms of quartz, feldspar and
lithics, QFL. Various percentages of QFL can
be plotted on a triangular graph as for
sandstones. The main division is between
feldspar-rich arkoses and lithic-rich wackes.

feldspar tend to be light. The dark
minerals separate out into what are
called mafic magmas, because they are
rich in MAgnesium and FerrIC (iron)
compounds. The light minerals are
concentrated in felsic magmas which
are rich in FELdspar and SIliCa
minerals. So mafic igneous rocks are
dark in colour; felsic rocks are light.
Because they are rich in silica, light-
coloured rocks are also said to be
silicic or acidic, while mafic rocks,
which are low in silica, like basalt, are
said to be basic.

Sedimentary rocks
This type of rock can form either
from fragments of weathered rock or
from minerals dissolved in water.
Rocks made from rock fragments or
'clasts' are called clastic rocks. Rocks
made from minerals dissolved in water
are called chemical rocks, or
biochemical rocks if they are made
from chemicals derived from living
things, such as the shells of shellfish.
     Clastic rocks include sandstones and
shales. They are made from fragments
of rock that don't dissolve in water;
those that do dissolve go on to form
chemical rocks. Insoluble fragments
are mostly silica-based minerals so
clastic rocks are sometimes called
siliclastic rocks.
     Siliclastic rocks are typically
classified according to the size of
particle they are mostly made of. They
mostly fall into three broad groups:

fine-grained lutites such as shale and
clay; medium-grained arenites
including many sandstones; and
coarse-grained rudites such as breccia
and conglomerate. But there are rocks
that don't fit so neatly into these
groups, such as wackes, sandstones
made from a mix of grains. So some
geologists prefer to classify according
to the proportion of the three main
grain types they contain – quartz sand,
feldspar and lithics (rock fragments),
sometimes known as QFL, plus the
fine 'matrix' material. Using QFL,
rocks such as sandstone are split into
feldspar-rich arkose sandstones and
lithic sandstones.
     Chemical and biochemical rocks
form into several groups. The biggest
is the carbonates, including limestones
and dolomite, made from calcium and
magnesium carbonate. Others are
chert, chalk, tufas and coal.

Metamorphic rocks
On a broad level, metamorphic rocks
are divided into granular or non-
foliated rocks and foliated rocks. With
the exception of hornfels, granular
rocks are made mostly from a single
mineral, such as marble from calcite
and quartzite from quartz. They are
formed mostly by the heat of close
contact with hot magma. Foliated
rocks are more complex. They are
characterized by their layered texture,
the result of the intense compressional
pressure brought about by regional scale

metamorphism. They are divided into
low-grade metamorphic rocks such as
slate; medium-grade rocks such as
schist and high-grade rocks such
as gneiss and granulite.
     This works as a basic classification,
but the composition of regionally
metamorphosed rocks is especially
complex and varied, so some geologists
look at metamorphic rocks in terms of
facies, conditions in which particular
assemblages of minerals are formed
(see Schists: Metamorphic facies,
World Directory of Rocks). Or they
might work in terms of the zones,
such as the Barrovian and Buchan
zones, in which particular rocks and
minerals are formed (see Gneiss and
Granulite: Mineral identifiers, World
Directory of Rocks).

Below: Gneiss's dark and light zebra stripes
of different minerals mark it out clearly as a
foliated metamorphic rock. Non-foliated rocks
show no such layered markings.

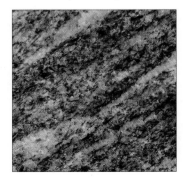

WORLD DIRECTORY OF ROCKS

The World Directory of Rocks gives detailed profiles of over 100 species of rock. Using the identification tips in the following pages, you can begin to gauge the type of rock you have recovered from the field, then follow up these clues in the Directory to draw a more acccurate conclusion.

Rocks are made of countless grains packed together – that is, they are aggregates. Sometimes the grains are as least as big as sugar crystals and clearly visible to the naked eye. Others are too tiny to see except under a microscope. A few, like obsidian, have no grains at all.

A few rocks (such as the conglomerates and breccias) are made from large fragments of other rocks. More often, though, the grains in rocks are mineral crystals. Some rocks are made from just a single mineral. Marbles, for instance, can be almost pure calcite. Most, however, are made of at least two or three different minerals such as eclogite which is garnet and augite, and granite which is mica, quartz and feldspar. Most rocks

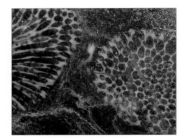

Above: A high fossil content often points to the sedimentary rock limestone, which forms from the remains of ancient marine creatures.

also contain a large number of 'accessory' minerals which are present in the rock but not in sufficient quantities to make a significant difference to its nature.

Rocks are typically divided into three major groups according to how they formed: igneous (from molten rock), sedimentary (from layers of sediment) and metamorphic (altered by extreme heat and pressure). This is how the specimens have been organized in the Directory.

Igneous rocks are further divided into extrusive (those that solidify from magma on the surface) and intrusive (those that solidify underground). Very loosely, the Directory section on Igneous rocks moves first through extrusive rocks from acidic, silica-rich or 'felsic' rocks like rhyolite, through intermediate rocks such as andesite to basic, silica-poor ('mafic' and 'ultramafic') rocks such as picrite. It then moves through intrusive rocks

IGNEOUS ROCKS

Rock name: may be plural if several varieties are profiled.

Identification: Notes the typical techniques used to identify specimens recovered from the field.

Data panel: Quick reference tool summarizing standard rock characteristics.

Norite

Norite is a similar rock to gabbro, based on a mix of plagioclase, pyroxene and olivine, and the two often form in the same large, layered intrusions as the mix separates during crystallization. Norite contains very slightly less plagioclase than gabbro, but the real difference is that gabbro's pyroxene is a clinopyroxene such as augite, while norite's is an orthopyroxene such as hypersthene. Unfortunately, the two can look so alike that they are impossible to distinguish without a microscope. Norite typically occurs in small, separate intrusions, or as layers along with other mafic igneous rocks such as gabbro. Norite also formed in association with ancient basalt intrusions, beneath huge basalt dyke swarms. One famous norite intrusion is at Sudbury in Ontario. Here a cavity 30m/98ft deep has been excavated from solid norite to house the Neutrino Observatory to detect neutrinos, minute particles streaming from the stars. Norite is unusually low in natural radioactivity and acts as a shield to allow scientists to block out unwanted background radiation.

Identification: Norite is a dark grey rock with a slightly matted look dominated by quite long, prismatic black hypersthene or enstatite crystals. It looks very like gabbro, but the plagioclase feldspar tends to be sandy coloured, while in gabbro it is whiter.

Grain size: Phaneritic (coarse-grained), occasionally pegmatitic
Texture: Even-grained or porphyritic
Structure: Layering and xenoliths are common
Colour: Dark grey, bronze
Composition: Silica (58%), Alumina (17%), Calcium and sodium oxides (10.5%), Iron and magnesium oxides (11%), Potassium oxides (2%)
Minerals: Plagioclase feldspar (labradorite or bytownite); Pyroxene (hypersthene); Olivine; A little hornblende, biotite mica, quartz and alkali feldspar
Accessories: Magnetite, apatite, ilmenite, picotite
Phenocrysts: Plagioclase feldspar, hornblende
Formation: Intrusive: dykes, stocks, bosses, often with gabbro
Notable occurrences: Aberdeen, Banff, Scotland; Norway; Great Dyke, Zimbabwe; Bushveld complex, S Africa; Sudbury, Ontario

Grain size; Texture: Grain size is the first step to identifying igneous rocks. Are grains visible to the naked eye? Texture is variations in grain size and shape.

Structure: Large-scale structures such as layers.

Colour: Overall colour impression.

Composition: This is the overall chemical content. Silica-rich rocks are generally lighter in colour.

Minerals: The major minerals that define the character of the rock.

Accessories: Any mineral not essential to the rock's character.

Phenocrysts: Unusually large crystal.

Formation: Where the magma solidifies to form the rock.

Notable occurrences: Entries are country by country except for Canada and the USA which are listed last, state by state. Each site in a country is separated by a comma. Additional location information is given in brackets.

Profile: The main features of the rock, its formation and characteristics.

Specimen photograph: Some important features may be annotated.

that form large intrusions such as granite, to those that form small intrusions such as pegmatites.

Sedimentary rocks are divided into clastic (formed from rock fragments), biogenic (formed by living things) and chemical (formed from once-dissolved chemicals). Clastic rocks are ordered from fine-grained lutites such as shale

and clay through medium-grained arenites (sandstones) to coarse-grained rudites (conglomerates and breccias).

Metamorphic rocks are divided according to whether they display any pressure flaking, banding or striping into non-foliated and foliated, but there is considerable overlap, partly because some rocks, such as

amphibolite, can be either foliated or non-foliated. Foliated rocks are ordered, very loosely, from those that are formed under least pressure (low-grade rocks such as slate and phyllite) to the most (high-grade rocks such as schist and gneiss). Rocks formed from meteorites such as tektites and suevites make an additional category.

SEDIMENTARY ROCKS

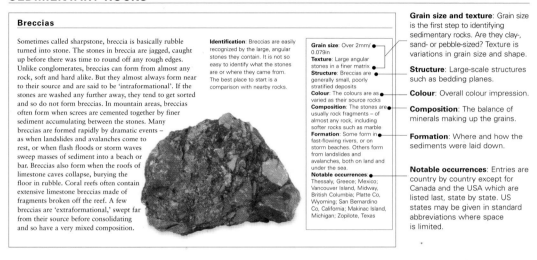

Breccias

Sometimes called sharpstone, breccia is basically rubble turned into stone. The stones in breccia are jagged, caught up before there was time to round off any rough edges. Unlike conglomerates, breccias can form from almost any rock, soft and hard alike. But they almost always form near to their source and are said to be 'intraformational'. If the stones are washed any further away, they tend to get sorted and so do not form breccias. In mountain areas, breccias often form when screes are cemented together by finer sediment accumulating between the stones. Many breccias are formed rapidly by dramatic events – as when landslides and avalanches come to rest, or when flash floods and storm waves sweep masses of sediment into a beach or bar. Breccias also form when the roofs of limestone caves collapse, burying the floor in rubble. Coral reefs often contain extensive limestone breccias made of fragments broken off the reef. A few breccias are 'extraformational,' swept far from their source before consolidating and so have a very mixed composition.

Identification: Breccias are easily recognized by the large, angular stones they contain. It is not so easy to identify what the stones are or where they came from. The best place to start is a comparison with nearby rocks.

Grain size: Over 2mm/0.079in
Texture: Large angular stones in a finer matrix
Structure: Breccias are generally small, poorly stratified deposits
Colour: The colours are as varied as their source rocks
Composition: The stones are usually rock fragments – of almost any rock, including softer rocks such as marble
Formation: Some form in fast-flowing rivers, or on storm beaches. Others form from landslides and avalanches, both on land and under the sea.
Notable occurrences: Thessaly, Greece; Mexico; Vancouver Island, Midway, British Columbia; Platte Co, Wyoming; San Bernardino Co, California; Makinac Island, Michigan; Zopilote, Texas

Grain size and texture: Grain size is the first step to identifying sedimentary rocks. Are they clay-, sand- or pebble-sized? Texture is variations in grain size and shape.

Structure: Large-scale structures such as bedding planes.

Colour: Overall colour impression.

Composition: The balance of minerals making up the grains.

Formation: Where and how the sediments were laid down.

Notable occurrences: Entries are country by country except for Canada and the USA which are listed last, state by state. US states may be given in standard abbreviations where space is limited.

METAMORPHIC ROCKS

Inset detail: Illustrations of related species and special features.

Mica schists

The most common schists are mica schists. Flakes of mica give them the most marked schistosity, and a real shine. The flakes are typically about 0.5mm/0.02in thick and can often be prised off with a knife. There is plenty of quartz in mica schist too, often concentrated in mica-poor layers, and a fair amount of albite feldspar. Sometimes red garnet or green chlorite crystals are visible. The mica in mica schists can be muscovite, sericite or biotite. Biotite is usually brown; muscovite and sericite are pale coloured and called white mica. If a white mica is fine-grained, it is called sericite; if it is coarser it is called muscovite. Most mica schists contain all three, but one usually predominates. Muscovite and sericite schists develop where metamorphism

Biotite schist: Biotite schist is brown and dark, but still has the mica gleam.

is of moderate intensity, and are often associated with greenschists and phyllites. Biotite develops partly at the expense of muscovite (and chlorite) when metamorphism becomes more intense. Even more intense metamorphism edges the schist towards garnet schists.

Muscovite schist: Muscovite schists, with sericite schists, are the lightest-coloured schists, coloured by white mica. Unlike sericite schist, the grains of mica in muscovite schist are clearly visible to the naked eye.

Mica (muscovite, sericite or biotite) schist
Rock type: Foliated, regional metamorphic
Texture: Medium to coarse-grained, sometimes with porphyroblasts. Thin schistose layering of mica flakes always marked.
Structure: Typically folded, on a small or large scale
Colour: Light grey, greenish (biotite schist is browner)
Composition: Quartz and mica (usually muscovite), plus kyanite, sillimanite, chlorite, graphite, garnet, staurolite
Protolith: Mainly mudrocks such as shale
Temperature: Moderate
Pressure: Moderate
Notable occurrences: Connemara, Ireland; Scotland; Scandinavia; Alps, Switzerland; Black Forest, Germany; Quebec; Duchess Co, NY; New Hampshire

Rock type: The degree of metamorphic banding and the mode of formation.

Texture: Overall grain size and variation.

Structure: Large-scale structures such as layers.

Colour: Overall colour impression.

Composition: The balance of minerals making up the grains.

Protolith: The original rock from which it was metamorphosed.

Temperature; Pressure: The conditions under which the rock was metamorphosed: low, medium or high-grade.

Notable occurrences: Entries are country by country except for Canada and the USA which are listed last, state by state.

IDENTIFYING ROCKS

Often dull greys, browns and blacks, rocks all look much the same at first glance. Yet they are as individual and distinctive as people, and once you know what to look for you will find it is quite easy to identify most of them and put them into their family groups.

Often the biggest clue to a rock's identity is where you found it. If you found it in an area of sandstone, it is likely to be sandstone. Before you begin to examine the specimen closely, see if you can spot formations of similar rock nearby that may yield useful clues. In cliffs and rock faces, the clear layering and beds of sedimentary rocks are usually unmistakable. Other rocks also create distinctive landscapes and landforms (see Rock Landscapes, Understanding How Rocks Are Made).

When you begin to examine your specimen in detail, the first task is to decide whether it is igneous, sedimentary or metamorphic.

Above: With a good eyeglass, you can often identify individual minerals within the rock, especially in medium- and coarse-grained igneous rocks such as granites.

Sedimentary rocks are pale in colour and tend to have similar grains, often held together by a cement. They may crumble as you rub them. Look for bedding planes and fossils.

Igneous rocks are identified by a harder, often shiny more compact look with a tightly packed interlocking mix of crystals. There should never be any layers or bands, except occasionally in layered granites and gabbros.

Some metamorphic rocks can look quite similar to igneous rocks, but foliation (layering and banding) is never found in igneous rocks. Granular metamorphic rocks tend to have a hard, shiny, sugary look and are more evenly dark or light, unlike igneous rocks which are quite often mottled.

The guides to identity given here are intended as a starting point, and should be used in conjunction with the clues given in the Directory.

Igneous rocks
COLOUR AND TEXTURE

	Light-coloured (silica-rich, acidic, felsic)	Medium-coloured (intermediate)	Medium-coloured (intermediate, feldspathic)	Dark-coloured (silica-poor, basic, mafic)
Fine-grained (aphanitic, volcanic, extrusive)	Rhyolite: White, grey, pink	Andesite: Salt and pepper (black and white)	Trachyte: Brownish-grey	Basalt: Dark grey to black
Medium-grained and porphyritic (hypabyssal, dyke, sill)	Quartz porphyry: White, grey, pink with light spots	Andesite porphyry: Dark grey, black with white spots	Monzonite: Dark grey with pale spots	Dolerite: Dark grey to black
Coarse-grained (phaneritic, plutonic, intrusive)	Granite: White, grey, pink; pinkish or whitish (tonalite')	Diorite: Salt and pepper (black and white)	Syenite: White, grey, pink; pinkish	Gabbro: Dark grey to black

OTHER TEXTURES AND COMPOSITIONS

Foam-like (vesicular) and glassy	Pumice: Whitish with fibrous look; very light	Scoria: Black to brown; very light	Vesicular basalt: Black to brown; heavy	Obsidian: Black, red, brown; glassy
Medium-to-coarse-grained carbonate and ultramafic rocks	Carbonatite: White with small grey spots	Dunite: Pale khaki to brown; ultramafic	Lamprophyre: Dark grey, with dark, shiny phenocrysts	Peridotite: Light to dark green; ultramafic

Metamorphic rocks

GRANULAR (NON-FOLIATED) ROCKS with no obvious layers, bands or stripes		FOLIATED ROCKS with obvious layers, bands or stripes	
Won't scratch glass	Will scratch glass	Grains often too small to see	Grains visible to the naked eye
Marble: Smooth feel; fizzes in dilute hydrochloric acid	Hornfels: Dark grey and black, dull, massive, conchoidal fracture	Slate: Dull grey, black, green, rings when struck; splits in thin sheets	Schist: Stripy bands or schistosity, platy cleavage
Dolomite marble: Smooth; powder fizzes in dilute hydrochloric acid	Metaquartzite: Pale translucent colours; fused quartz grains	Phyllite: Shiny grey, black, green; splits in thin sheets; may be striped	Biotite mica schist: dark; pale schist is muscovite mica
Greenstone: Greenish; harder than a fingernail unlike soapstone	Eclogite: Pale green pyroxene with red garnet	Mylonite: Streaky, smeared out texture	Gneiss: Tough, minerals separated into light and dark bands
Serpentinite: Greasy feel; green, yellow, brown or black	Amphibolite: Black, shiny crystals of amphiboles; also foliated	Glaucophane/Blue schist: Bluish colour, slender fibrous crystals	Granulite: No black mica, lenses of pale quartz and feldspar

Sedimentary rocks

No visible grains	Sand-size, visible grains	At least gravel-sized grains	Biochemical rocks that react with vinegar if powdered
Siltstone: Gritty feel; hard enough to scratch glass	Sandstone: Even, sand-sized grains; may be yellow, brown or red	Breccia: Large angular fragments set in mudlike mix	Chalk: White, powdery, leaves white mark, feels gritty
Claystone: Smooth feel; too soft to scratch glass	Ironstone: Even, sand-sized grains; very dark brown, red	Conglomerate: Large round pebbles set in mudlike mix	Oolitic limestone: Buff-coloured, tiny spheres like fish roe
Shale: Smooth feel; too soft to scratch glass	Greensand: Even, sand-sized grains; greenish colour	Boulder clay: Huge mix of stones set in muddy clay	Pisolitic limestone: Sand-coloured, tiny spheres like small peas
Marl: Earthy, slimy feel; too soft to scratch glass	Arkose: Even, sand-sized grains; won't crumble; pinkish, red		Fossil limestone: Pale grey, packed with fossils
Chert: Smooth, hard and glassy appearance	Greywacke: Mix of sand and fragments of rock		Dolomite: Dull grey; weathers pink or brown

VOLCANIC ROCKS: SILICA-RICH ROCKS

Volcanic rocks are formed mainly when lava erupting from volcanoes cools and solidifies, but any material ejected from a volcano can form volcanic rock if it turns to stone, including ash, blobs of molten rock and froth. Because exposure to the air cools lava quickly, before crystals have time to grow, volcanic rocks are usually aphanatic (fine-grained) or even glassy. Volcanic rocks that are rich in silica (at least 55 per cent) are light in colour and include rhyolite, quartz porphyry and dacite.

Rhyolite

One of the most widespread volcanic rocks, rhyolite forms from the same silica-rich (70–78 per cent) magma that forms granite when it solidifies underground. This is the magma that melts its way up through the continental crust, so rhyolite is a continental rock. Rhyolite has been found on islands far from land, but such oceanic occurrences are rare. Rhyolite is the fine-grained, extrusive equivalent of granite, but there are subtle differences in chemistry. The mica in rhyolite is black biotite, not the brown muscovite seen in granite, and its potassium feldspar is sanidine while that of granite is orthoclase.

The high quartz content of rhyolite magmas means they are relatively cool and very viscous, and this sticky magma tends to clog up the volcanic vent. Sometimes, a plug is left behind long after the volcano has died, leaving a spire of rhyolite as it is gradually exposed by weathering. More often, the plug is blasted away in a mighty explosive eruption, which is why rhyolite is linked to some of the world's most explosive volcanoes, especially caldera complexes such as Tambora in Indonesia. Explosive eruptions are fuelled by the sudden expansion of steam and carbon dioxide in the magma. The explosion blasts away fragments of the plug in huge clouds of ash and deadly avalanches of pyroclasts. Gas bubbles turn parts of the lava to a froth that later solidifies as pumice. So rhyolite rock often forms from ash and pyroclasts rather than lava. Only once there is less gas in the magma can the rhyolite flow on to the surface as lava. Rhyolite lava piles up thickly in domes or coulées (tongues) around the vent – too sluggish to flow far. Rhyolite flows have broken, blocky surfaces, because the rind shatters as the inner mass creeps forward. Ancient rhyolites may have flowed farther, though, if, as some geologists argue, they were superheated and made less viscous by hotspots.

Because they erupt on the surface and cool rapidly, rhyolites are basically fine-grained. Indeed, where they have been quenched (cooled ultra-quickly), they are mostly glassy. Glassy rhyolites include obsidian, pitchstone and perlite. Yet because rhyolites are so viscous, they usually contain phenocrysts – large crystals that formed while the magma lingered in the volcano's magma chamber. Sometimes phenocrysts dominate so much that rhyolite can look like granite, with the microcrystalline groundmass visible only under a microscope. Rocks like these are called nevadites.

Banded rhyolite (below):
When rhyolitic lava erupts on the surface, smaller crystals often respond to the flow by aligning themselves in bands, an effect called flow-banding. Banded rhyolite is sometimes called wonderstone, and valued by collectors – especially when it contains cavities filled with silica precipitates such as agate.

Spherulitic rhyolite (above):
Sticky, rhyolitic lava often traps pockets of volatile vapours. In quickly cooled glassy rhyolites, some gas pockets develop into spherulites – balls of radiating needle-like crystals of quartz and feldspar – forming spherulitic rhyolite. Spherulites are typically a few millimetres across, but can be up to a metre.

Grain size: Aphanitic (fine-grained)
Texture: Phenocrysts common; alternating layers of grains common; flow-banding common (Banded rhyolite)
Structure: Vesicles and other remnant bubbles common; may contain spherules (Spherulitic rhyolite)
Colour: Usually light-coloured – pinkish or reddish brown, but also white, greenish, grey
Composition: As for granite: Silica (74% average), Alumina (13.5%), Calcium and sodium oxides (less than 5%), Iron and magnesium oxides (less than 3.5%)
Minerals: Quartz; Potassium feldspar (sanidine) and plagioclase feldspar (oligoclase); Biotite mica
Accessories: Aegerine, zircon, apatite, magnetite, amphibole or pyroxene
Phenocrysts: Quartz, orthoclase and oligoclase feldspar, hornblende, biotite mica, augite
Formation: Lava flows, dykes, volcanic plugs in continents
Notable occurrences: Lake District, Shropshire, England; Snowdonia, Wales; Vosges, France; Black Forest, Saxony, Germany; Carpathian Mountains, Austria; Siebenbürgen, Romania; Tuscany, Italy; Iceland; Caucasus, Georgia; Rocky Mountains including Yellowstone; Arizona.

Rhyolite volcanoes include: Tambora, Indonesia; Mount Kilimanjaro, Kenya/Tanzania; Yellowstone, Wyoming; Crater Lake, Oregon.

Quartz porphyry

Quartz porphyry is a loose term for a rock with a similar chemical composition to granite and rhyolite that contains phenocrysts of white quartz (or, less often, orthoclase feldspar) that spot the rock like chunks of fat in a burger. The basic matrix of crystals around the phenocrysts is usually fine-grained, like rhyolite, because the melt containing the phenocrysts was fed into narrow dykes where it cooled and solidified quickly. So the average grain size is akin to granite. More recent quartz porphyries are dyke rocks, but ancient formations that formed in the Palaeozoic age of Earth's history – over 550 million years ago – were lava flows. So some geologists prefer to call ancient quartz porphyries palaeorhyolite. Many of these ancient quartz porphyries have been crushed and sheared by earth movements in their long history, giving them a striped look like schists. When the phenocrysts have been preserved, these rocks are called porphyry schists, and American geologists sometimes call them aporhyolite. They are well known from the Swiss Alps and from England's Charnwood Forest. The supposedly metamorphic halleflintas of Scandinavia may also have been formed like this.

Identification: Quartz porphyry is easily identified by the large white or grey blobs of quartz and feldspar in the reddish brown matrix of a rhyolitic mix of fine-grained crystals or even glass.

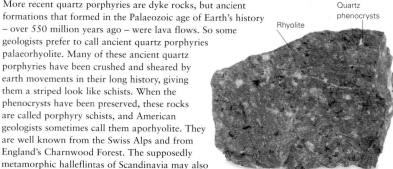

Rhyolite

Quartz phenocrysts

Grain size: Mixed
Texture: Phenocrysts in fine-grained, microcrystalline or glassy matrix
Structure: Vesicles rare
Colour: Usually light-coloured – red, brown, greenish
Composition: As for granite: Silica (74% average), Alumina (13.5%), Calcium and sodium oxides (< 5%), Iron and magnesium oxides (< 4%)
Minerals: Quartz; Potassium feldspar (sanidine) and plagioclase feldspar (oligoclase); Biotite mica
Accessories: Hornblende, augite, bronzite, garnet, cordierite, muscovite
Phenocrysts: Quartz, orthoclase feldspar
Formation: Dykes, or ancient lavas in continents
Notable occurrences: Devon, Cornwall, Charnwood Forest (Leics), England; Westphalia, Germany; Alps, Switzerland; San Bernardino Co, CA; Lake and St. Louis Co, MN; Green Lake Co, Wisconsin; Pennsylvania.

Porphyries
Porphyries are igneous rocks that contain large, conspicuous crystals in a groundmass of finer crystals. The phenocrysts, as the larger crystals are called, formed early on in the middle of the molten magma; the finer crystals formed later, typically after the magma containing them erupted or was injected into a dyke. In the micrograph of trachybasalt shown above, for example, large phenocrysts of olivine, clino-pyroxene (the large twinned crystal) and plagioclase feldspar are set in a fine-grained groundmass of the same crystals. The word porphyry now applies to any igneous rock with phenocrysts, but originally it referred to the beautiful red porphyries used by the Ancient Egyptians and the Romans in the time of the emperor Claudius. This rock, which the Romans called *porfido rosso antico*, was taken from a dyke 30m/98ft 5in thick on the Red Sea at Jebel Dhokan, and contained white or rose red pheno-crysts of plagioclase, dark black hornblende and plates of iron oxide, all in a dark red groundmass.

Dacite

Named after the Roman province of Dacia in modern Romania, dacite can be a beautiful rock when polished, but is commonly used for road-chippings. It forms from fairly viscous lava (55–65 per cent silica) in lava flows and dykes. It can also form massive intrusions in the heart of old volcanoes, creating lava domes such as Mount St Helens in Washington. It contains less quartz than rhyolite and so is intermediate in composition between rhyolite and the basic lava andesite. The quartz is often in the form of rounded phenocrysts in the groundmass, a little like quartz porphyry. Dacite also has andesine and labradorite as its feldspars rather than sanidine as rhyolite.

Biotite, hornblende

Feldspar

Grain size: Aphanitic (fine-grained) or even glassy
Texture: Phenocrysts, alternating layers of grains and flow-banding common
Structure: Vesicles and other remnant bubbles common
Colour: Usually light-coloured – reddish or greenish
Composition: Silica (65% av), Alumina (16%), Calcium & sodium oxides (8%), Iron & magnesium oxides (6.5%)
Minerals: Quartz; Potassium feldspar (andesine and labradorine) and plagioclase feldspar; Biotite; Hornblende
Accessories: Pyroxene (augite and enstatite), hornblende, biotite, zircon, apatite, magnetite
Phenocrysts: Quartz, feldspar, hornblende, biotite
Formation: Lava flows, dykes
Notable occurrences: Argyll, Scotland; Massif C, France; Saar-Nahe, Germany; Hungary; Siebenbürgen, Romania; Almeria, Spain; New Zealand; Martinique; Andes, Peru; Rocky Mountains; Nevada.

Identification: Dacite's rich hornblende and biotite are grey or yellowish, with white specks of feldspar like this. Augite and enstatite-rich dacites are darker.

ANDESITES

Midway between rhyolite and basalt in silica content, andesites are the most common volcanic rocks after basalt and are found all around the world near subduction zones. They get their name from the Andes in South America, and are associated with all the classic cone-shaped volcanoes, such as Mount Fuji in Japan and Mount Edgecumbe in New Zealand.

Andesite

Identification: Most andesite has a classic 'salt and pepper' look with white tablet-shaped grains of plagioclase feldspar visible to the naked eye set in a dark, often almost black, groundmass of fine-grained, occasionally glassy, minerals – mainly biotite mica, hornblende and pyroxene. Typically the dark minerals make up about 40 per cent of the rock by volume, much less than in basalt.

Andesites are found pretty much anywhere that an oceanic plate is subducted beneath a continent. Here they create a string of volcanoes along the continental margin, or an arc of volcanic islands along the edge of the continental shelf. Andesites are especially common in areas of recent mountain building. Not only the Andes, but the entire cordillera of mountains running from the Andes to the Rockies is predominantly andesite. In fact, wherever there are volcanoes in the 'Ring of Fire' around the Pacific, there is likely to be andesite.

Continental and island arc volcanoes spew out mainly andesite, dacite or rhyolite, depending on their silica content, while oceanic volcanoes emit basic lavas such as olivine, basalt and trachyte. Geologists call the line separating the two the andesite line. It runs roughly down the west coast of the Americas, and down from Japan to New Zealand via the Marianas, the Palaus, the Bismarcks, Fiji and Tonga. Although not as silica-rich as rhyolite, is still fairly viscous and often gets clogged up in the vent of the volcano. Consequently, andesite generates some of the world's most dramatic eruptions, as pressure builds up high enough to blast through the plug.

Andesitic volcanoes frequently produce devastating pyroclastic volcanoes, and most andesite volcanoes are stratovolcanoes – the classic cone-shaped peaks in which layers of lava alternate with layers of ash and pyroclasts, as lava pours out after the ash in each eruption.

Andesite lava flows better than rhyolite, but still sluggishly. When erupted, it first forms a mound around the vent. This creeps down the flanks of the volcano, advancing barely a few metres a day. The lava moves so slowly that the outside of the flow cools and solidifies. So, as it moves, the surface of the flow breaks up into a jumble of angular blocks that look like rubble. Even the hottest, most fluid andesites rarely flow farther than 10km/6 miles from the vent.

Porphyritic andesite: Many andesites are porphyritic – that is, they are spattered with large grains, clearly visible to the naked eye, that formed before the lava erupted. When these phenocrysts are especially large, the rock is called porphyritic andesite. The phenocrysts are typically plagioclase feldspar (white), but can be pyroxenes and amphiboles (usually greenish black).

Grain size: Aphanitic (fine-grained), occasionally glassy
Texture: Often porphyritic
Structure: Often displays flow structure; occasional vesicles
Colour: Grey, purplish, brown, green, almost black
Composition: Silica (59% average), Alumina (17%), Calcium and sodium oxides (10%), Iron and magnesium oxides (11%)
Minerals: Quartz; Feldspar: plagioclase feldspar (andesine) plus small amounts of potassium feldspar (oligoclase, sanidine); Biotite mica; Amphibole (hornblende); Pyroxene (augite)
Accessories: Magnetite, apatite, zircon, olivine
Phenocrysts: Plagioclase feldspar, pyroxenes such as augite, amphibole, hornblende
Formation: Extrusive lavas, ashes and tuffs in subduction zones and areas of mountain building
Notable occurrences: Glencoe, Scotland; Lake District, England; Snowdonia, Wales; Vosges, Auvergnes, France; Rhineland, Germany; Siebenbürgen, Romania; Caucasus, Georgia; Andes; Rocky Mountains. Andesite volcanoes include: Mount Fuji, Bandai-san, Japan; Krakatoa, Indonesia; Pinatubo, Philippines; Ngauruhoe, Mount Edgecumbe, Ruapehe, New Zealand; Citlaltépetl, Popocatépetl, Mexico; Mount Pelée, Martinique; Soufriere, St Vincent; Mount Shasta, California; Mount Hood, Oregon; Mount Adams, Washington.

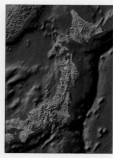

the plate may reflect the curvature of the Earth. There are many of these island arcs in the Pacific, including the Aleutians, the islands of Japan, the Marianas, Tonga and the Solomon Islands. The Antilles in the Caribbean also form an island arc. All these islands are volcanic and form when the subducted plate plunges down into the mantle, creating deep ocean trenches all along the margins of the plate. This satellite image of Japan (above) shows the Japan Trench, the dark area to the right of the islands, which forms part of the boundary between the Pacific and Eurasian plates. As the subducted plate sinks into the mantle, it melts and, in a complex process, forms magmas, including andesite, basalt and boninite. These hot magmas punch through the edge of the overriding plate like a needle stitching a hem, and as they penetrate, they erupt on the surface as volcanoes.

Boninite

This rarity takes its name from the Izu-Bonin-Mariana chain of islands south of Japan in the Pacific. Most of the boninites formed here between about 30 and 50 million years ago, but boninites are still forming today. They are associated almost exclusively with island arcs, and seem to require particular conditions for their formation. They are formed when an ocean tectonic plate is subducted beneath another oceanic plate. The subducted plate carries sea water down into the mantle with it as it plunges into the Earth, and the water alters the chemistry of the magma formed as the plate melts in the heat of the mantle. Geologists suggest that boninites form only if high temperatures are reached fairly quickly as the plate is subducted – otherwise, andesites will form. They may be linked to the early stage of subduction.

Grain size: Glassy
Texture: Often porphyritic
Structure: Often displays flow structure; occasional vesicles
Colour: Dark grey, often black
Composition: Silica (59% average), Alumina (17%), Calcium and sodium oxides (10%), Iron and magnesium oxides (11%)
Minerals: Quartz; Feldspar: plagioclase feldspar; Pyroxene (augite, bronzite); Amphibole (hornblende)
Accessories: Pentlandite, spinel
Phenocrysts (small): Pyroxenes: augite, bronzite
Formation: Extrusive lavas, dykes and sills in island arcs
Notable occurrences: Crimea, Ukraine; Izu-Bonin-Marianas island chain, Pacific; North Tonga Ridge, New Hebrides; Setouchi, Japan. Possible continental margin occurrences: Isua, Greenland; Yukon, Canada; Glenelg, South Australia: Antilles, Caribbean.

Identification: Boninite is dark in colour, with small black phenocrysts set against a dark glassy groundmass.

Pyroxene andesite

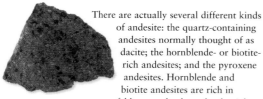

There are actually several different kinds of andesite: the quartz-containing andesites normally thought of as dacite; the hornblende- or biotite-rich andesites; and the pyroxene andesites. Hornblende and biotite andesites are rich in feldspar and coloured pale pink, yellow or grey. The pyroxene andesites are by far the most common of the andesites and occur almost as widely as basalt, and often come from the same magma source. The pyroxene in pyroxene andesite is usually augite – giving augite andesite – but can be olivine. The augite gives these andesites a sparkle when the rock is broken, but they are frequently altered to hornblende. Sometimes, good-sized augite crystals can be found in pyroxene andesite tuffs – that is, deposits of ash and pyroclasts.

Biotite andesite: Biotite andesite is typically yellow, pinkish or grey, often with black phenocrysts of amphibole or pyroxene.

Identification: Pyroxene andesite can look quite like basalt, and its high pyroxene content means that it is chemically closer too. But pyroxene andesite usually contains phenocrysts and is usually slightly lighter in colour, with traces of white feldspar.

Grain size: Aphanitic (fine-grained), occasionally glassy
Texture: Often porphyritic
Structure: Often displays flow structure; occasional vesicles
Colour: Grey, green, almost black
Composition: Silica (59% average), Alumina (17%), Calcium and sodium oxides (10%), Iron and magnesium oxides (11%)
Minerals: Quartz; Feldspar: plagioclase feldspar (andesine) plus small amounts of potassium feldspar (oligoclase, sanidine); Biotite mica; Amphibole (hornblende) or Pyroxene
Accessories: Magnetite, apatite, zircon, olivine
Phenocrysts: Pyroxenes such as augite and olivine, amphibole, hornblende
Formation: Extrusive lavas, ashes and tuffs in subduction zones and areas of mountain building
Notable occurrences: See andesite

TRACHYTES and SPILLITE

Trachytes and phonolites are medium-coloured, fine-grained volcanic rocks that flow easily enough to form lavas. They occur in many of the same places as basalts, including rifts, but contain more lighter-coloured minerals, and a modicum of quartz. They are all alkaline rocks, containing sodium and potassium feldspars. Trachyte is the mildest alkaline, phonolite is the strongest.

Trachyte

This volcanic rock is medium-coloured and very fine-grained. It is the volcanic equivalent of syenite, and erupts along rifts, in oceanic settings, above hot spots and in back arc basins between island arcs and the continental land mass. Trachytes are often associated with basalt, and are sometimes thought to be characteristic of a volcano past its prime. The parasitic cones of the Hawaiian shield volcanoes often ooze trachyte lava, for instance. Yet trachytes can also form quite extensive lava flows in their own right, as they do in Saudi Arabia.

Trachyte is fine-grained, but unlike andesite and rhyolite, very rarely glassy. It seems that crystals nearly always form, even though they may be microscopically small. Remarkably, rectangular phenocrysts of white sanidine are often already formed in the lava when it erupts. These tend to line up with the direction of the lava flow, and through a microscope you can often see that the smaller crystals form flow patterns around them. This texture is described as trachytic, and is sometimes found in other lavas, such as Hawaiite.

Although they are too small to see, trachyte is full of tiny cavities left by gas bubbles. It is these that give the rock the slightly rough feel to which it owes its name, from *trachys*, the Greek for 'rough'. These cavities sometimes fill with tiny crystals of the silica minerals tridymite, cristobalite, opal and chalcedony. The bulk of trachyte, though, is mostly alkali (sodium- and potassium-rich) feldspars, notably sanidine (in rodlike microcrystals as well as phenocrysts), along with dark-coloured minerals such as biotite, amphiboles (hornblende) or pyroxenes such as aegerine and diopside. An increase in the silica content takes trachyte towards rhyolite; a decrease, with a corresponding increase in feldspathoids such as leucite, nepheline and sodalite, takes it towards phonolite.

Identification: Trachyte is a brownish grey, medium-coloured rock, which is microcrystalline but almost never glassy. The dark groundmass is usually spotted with thin white phenocrysts of sanidine. A tell-tale clue to its identity is the rough feel created by tiny gas bubbles.

Grain size: Aphanitic (fine-grained), can occasionally be glassy
Texture: Even, but often porphyritic
Structure: Often displays flow structure called trachytic visible only under a microscope in which crystals of groundmass appear to flow around phenocrysts. Some specimens also have minute steam cavities, making the surface of the rock feel rough.
Colour: Usually grey, but can be white, pink or yellowish
Composition: Silica (62% average), Alumina (17%), Calcium and sodium oxides (8%), Iron and magnesium oxides (6%)
Minerals: Quartz; Feldspar: potassium feldspar (sanidine) and plagioclase feldspar (oligoclase); Biotite mica; Amphibole (hornblende, often altered to magnetite and augite); Pyroxene (aegerine, diopside)
Accessories: Apatite, zircon, magnetite, leucite, nepheline, sodalite, analcime. Plus in cavities: tridymite, cristobalite, opal, chalcedony.
Phenocrysts: Tablet-shaped sanidine, often aligned with the flow
Formation: Extrusive lavas, dykes and sills often in association with basalt
Notable occurrences: Skye, Midland Valley, Scotland; Lundy Island, Devon, England; Eifel, Thuringia, Saar, Berkum, Drachenfels (Rhineland), Germany; Auvergne, France; Naples, Ischia, Sardinia, Italy; Iceland; Azores; Saudi Arabia; Ethiopia; Madagascar; Cambewarra (New South Wales), Australia; Hawaii; Black Hills, South Dakota; Colorado

Sanidine

Porphyritic trachyte: Trachyte is basically a medium-coloured rock, with a groundmass of dark minerals such as biotite, hornblende and pyroxenes, and light-coloured sanidine feldspar. Interestingly, the sanidine forms in two stages, and the rock may be spotted with large long white sanidine phenocrysts that formed early in the magma. Under a microscope, trachytic flow patterns are visible in the small crystals around them.

Phonolite

Phonolites are mostly quite recent rocks, all forming within the Tertiary Age – that is, in the last 66 million years. They are medium-coloured, fine-grained volcanic rocks. They split into thin slabs and have such a compact structure that the slabs ring when struck with a hammer, which is why they were once called clinkstone. Phonolites are quite similar to trachyte, and the two occur in similar places. But phonolites are richer in alkaline minerals. Their lower silica content favours the formation of feldspathoids such as nepheline, leucite and sodalite, rather than the potassium feldspars of trachyte. So phonolites are the extrusive equivalent of nepheline syenite, rather than plain syenite. Like trachyte, they contain two generations of crystals. The first generation are the large, flattened tablet-shaped crystals of sanidine and nepheline, which form slowly in the magma. These become phenocrysts when the lava is erupted, and smaller crystals quickly form around them, often with a microscopic trachytic flow structure. Sometimes, leucite replaces nepheline to create leucite phonolite, as found near Naples, which is often studded with blue hauyne crystals, and sphene.

Identification: Phonolite is usually a mottled grey, but tiny needles of pyroxene can turn it greenish. The way phonolite breaks into flat slabs is often a real clue to identity, especially if the slabs give a metallic clink when hit with a hammer.

Grain size: Aphanitic (fine-grained), occasionally glassy
Texture: Dense, porphyritic
Structure: Platy structure, so breaks into slabs
Colour: Dark green, grey
Composition: Silica (57.5% av), Alumina (19.5%), Calcium & sodium oxides (11%), Iron & magnesium oxides (6%)
Minerals: Alkali feldspar (sanidine, anorthoclase); Foids (nepheline, leucite, sodalite, hauyne, nosean); Pyroxene (aegerine, diopside); Amphibole (barkevikite hornbl'd, riebeckite)
Accessories: Apatite, zircon, magnetite, sphene, garnet
Phenocrysts: Sanidine, nepheline, aegerine
Formation: Extrusive lavas, dykes and sills often with trachyte and nepheline syenite
Notable occurrences: Wolf Rock (Cornwall), England; Eildon, Scotland; Auvergne, France; Eifel, Laacher S, Germany; Bohemia, Czech Rep; Naples, Sardinia, Italy; Canary I; Cape Verde I; NSW; Cripple Creek, CO; Black Hills, SD; Devil's Tower, WY; Mt Erebus, Antarctica

Great East African Rift Valley
In few places is the power of tectonic plate movement more striking than in Africa's Great Rift Valley. The Valley is part of a huge set of fissures in Earth's crust called the East African Rift system, which threatens to split Africa in two. It started to open up about 100 million years ago, as plates on either side began to pull apart. As the crust stretched, volcanoes repeatedly burst through, and now dot the whole valley, including Erte Ale in Ethiopia and Ol Doinyo Lengai in Tanzania. Initially there were floods of basalt, then shield volcanoes emitting rhyolites and basanites, and finally volcanoes erupting trachyte and phonolite. The valley that formed is now over 6,000km/3,728 miles long and 50km/31 miles wide on average. The walls typically rise 900m/2,953ft above the valley floor, but at Mau in Kenya, cliffs soar to a height of 2,700m/8,858ft. Geologists believe that the Afar Triangle in Ethiopia, where the branches of the rift system meet, is the start of the world's next great ocean.

Spillite

Spillite is a medium-dark greenish black volcanic rock made from a groundmass of dark amphiboles such as actinolite and riebeckite, with occasionally bright cream phenocrysts of albite. It erupts mainly in oceanic locations, along with basalt, though it can often be found in ancient locations on land. It is one of the main kinds of magma that form pillow lavas. Pillow lavas are balls or tubes of lava that form where lava erupts slowly on the ocean bed. As the lava oozes up, contact with cold sea water quickly chills it to form a thin crust, and as the lava goes on pushing up it solidifies into a ball or tube, just like blobs of toothpaste squeezed from a tube.

Grain size: Aphanitic (very fine-grained)
Texture: Often porphyritic
Structure: Platy structure, so breaks into slabs
Colour: Dark green, black
Composition: Silica (50% average), Alumina (16%), Calcium and sodium oxides (13%), Iron and magnesium oxides (18%)
Minerals: Plagioclase feldspar (albite); Amphibole (actinolite, riebeckite); Chlorite; Epidote
Accessories: Apatite, zircon, magnetite
Phenocrysts: Albite, actinolite
Formation: Pillow lavas and tuffs
Notable occurrences: Oceanic crust (worldwide); Cornwall, England; Alaska; California

Identification: Spillite is dark, fine-grained volcanic rock quite similar to basalt. It can sometimes be identified by the way it breaks into slabs or by its formation as pillow lavas.

BASALTS

Black and fine-grained, basalt is the classic 'mafic' volcanic rock – rich in iron and magnesium minerals, and very low in silica. It is one of the earliest lavas to erupt from any volcano, coming straight up from the mantle in huge quantities, very hot and very fluid, and uncontaminated by the silicas that make other lavas much more viscous (less fluid).

Basalt

Alkali olivine basalt: The typical basalt rock is dark in colour with no visible grain structure. It tends to be black when freshly exposed, but turns reddish or greenish when weathered. Alkali or olivine basalt contains lots of olivine and augite in the groundmass (not as phenocrysts).

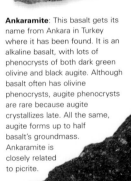

Ankaramite: This basalt gets its name from Ankara in Turkey where it has been found. It is an alkaline basalt, with lots of phenocrysts of both dark green olivine and black augite. Although basalt often has olivine phenocrysts, augite phenocrysts are rare because augite crystallizes late. All the same, augite forms up to half basalt's groundmass. Ankaramite is closely related to picrite.

Basalt is the most common rock on Earth. Much of it, though, is hidden away under the sea, for it forms the bulk of the ocean floor, which itself makes up 70 per cent of the Earth's surface. Basalt lava wells up through fissures in the ocean floor at the mid-ocean ridge as the two halves of the ocean floor pull apart. The lava freezes on to the receding edges of the two halves and ensures new rock is added as the ocean spreads away. It often forms pillow lavas here, as hot lava is suddenly chilled by the cold sea water to create countless cushion-like knobs of solid rock.

Basalt lava also wells up where hot spots penetrate the ocean floor to create island volcanoes such as Mauna Loa and Kilauea in Hawaii. Sometimes the hot basalt lava flows straight into the sea and is shattered as it suddenly freezes, and the fragments create black beach sands like those of Hawaii. Not all basalt is oceanic, though. Some basalt can erupt in continental fissures in a process no one quite understands. For example, huge floods of basalt have poured from fissures on to the surface to form gigantic plateaus, such as India's Deccan and North America's Columbia River Plateau. When it erupts, basalt lava is so hot and fluid that it can flow for tens of kilometres from the vent. When particularly hot it can even flow 500km/ 310 miles or so. This is why basalt habitually forms broad shield volcanoes, or flood basalt plateaux.

The shape the lava takes as it freezes depends on its temperature and its speed. When it is quite warm and fluid, surfaces wrinkle into rope-like ridges known by their Hawaiian name of *pahoehoe* (pronounced 'pa-hoy-hoy'). If they are a little more viscous or cooler, the surface tends to freeze and then break up into a jumble of rubble, a flow known as *aa* (pronounced 'ah-ah').

Grain size: Aphanitic (very fine-grained) or tachylytic (glassy)
Texture: Usually dense, with no visible mineral grains
Structure: Often porphyritic, and tends to include xenoliths (large lumps of other minerals) of olivine and pyroxene. Frequently spongy with vesicles or amygdaloidal cavities. Large masses of basalt may be cracked into hexagonal columns, like the Giant's Causeway (N. Ireland)
Colour: Black or blackish grey when fresh, may weather to reddish or greenish crust
Composition: Silica (50% average), Alumina (16%), Calcium and sodium oxides (13%), Iron and magnesium oxides (18%)
Minerals: Plagioclase feldspar (labradorite); Pyroxene; Olivine; Magnetite; Ilmenite. Tholeitic basalts (low in olivine): Plagioclase, pyroxene (hypersthene, pigeonite); Magnetite. Alkali basalts; (olivine-rich): Olivine, Pyroxene (augite); Magnetite.
Accessories: Countless
Phenocrysts: Green glassy olivine or black shiny pyroxene, or occasionally white tabular plagioclase feldspar
Amygdales: Zeolites, carbonates, and silica in the form of chalcedony and agate
Formation: Extrusive lavas, dykes and sills. Most basalts occur as lava flows from volcanoes or as sheets building up lava plateaux in flood basalts. Surfaces are either smooth, ropey pahoehoe or clinkery aa. Under the ocean basalt is often in balloon-like masses of pillow lava
Notable occurences: See opposite page

The Giant's Causeway
In the last stages of cooling, basalt lava flows often contract and fracture into extraordinary hexagonal columns. The most famous example of these is the Giant's Causeway in Antrim, Northern Ireland. Legend has it the stones were laid by the giant Finn MacCool to reach his lover in Scotland, but in fact they are a natural feature of basalt rock. About 65 million years ago, North America began to split apart from Europe, and basalt lavas welled up into the rift created. Here in Antrim, tholeitic lavas erupted over hundreds of thousands of years then stopped – only to start again abruptly. The lava poured into valleys and became so deeply ponded at the site of the Causeway that it formed a lava lake so deep that it cooled only slowly. As it slowly solidified and contracted, it developed regular hexagonal stress patterns. Soon six-sided columns 30–40cm/12–16in across permeated the whole cooling mass. Further eruptions followed, but this was the one that left its distinctive legacy.

Vesicular and amygdaloidal basalt

The grains in basalt are usually so fine, they are invisible to the naked eye, and the impression of the rock is a black mass, created by a mix of dark minerals – essentially labradorite (plagioclase), pyroxene and olivine, with magnetite and ilmenite. But it can be porphyritic, or contain cavities called vesicles that form as gas bubbles expanded in the solidifying lava. Larger vesicles seem to form more commonly in basalt pahoehoe than any other lava. It may be because the lava is so hot and fluid enough for the bubbles to expand easily. Unlike vesicles in other, more viscous lavas, those in pahoehoe are pretty much round rather than long. Rock with lots of empty vesicles like these is called vesicular basalt. Once the lava is set, water percolating through the lava often begins to fill many of these cavities with mineral crystals. These infillings are called amygdales, from the Greek for almond for their typical shape. They can be anything from 1mm to 30cm/0.04 to 12in across and typically contain quartz, carbonates and zeolites. Basalt full of amygdales is called amygdaloidal basalt.

Vesicular basalt: Basalt is often filled with countless vesicles, which make it look as if some insects have been at it. Other igneous rocks do have vesicles, but they are especially numerous and rounded in basalt. A dark black groundmass indicates that a rock full of little holes like these is vesicular basalt.

Typical rounded vesicles

Alkali and tholeitic basalt

There is a wide spectrum of basalt rocks, but they can be divided into two broad groups according to their chemistry – the tholeitic basalts and alkali basalts. Tholeitic basalts are low in olivine and rich in calcium-poor pyroxenes such as hypersthene and pigeonite. Most basalts that ooze up from rifts and mid-ocean fissures are tholeitic. So ocean floors are tholeitic basalt. Flood basalts are tholeitic too. Alkali basalts contain more sodium and potassium and are rich in olivine, the feldspathoid nepheline and calcium-rich pyroxenes such as augite. Geologists believe they are more alkaline because minerals in the magma have been less divided by partial melting. In the Hawaiian hotspot volcanoes, lavas that build the initial undersea cone are alkali basalts, cooled quickly by seawater. But as the cone rises above the sea, streams of fluid tholeitic lava spew out to create a huge shield, cooling slower in the air and more affected by partial melting. As the volcano dies down again, spurts of alkali lavas resume, creating a cap of alkali basalt rock. In the million-year life cycle of Mauna Kea, Hawaii's biggest volcano, this last subdued alkaline stage has gone on for 40,000 years, and may yet last another 60,000 years.

Notable occurrences:
Tholeitic basalts: Deccan, India; Red Sea; Paraná Basin, South America; Palisades, New Jersey; Rio Grande Rift, Mexico; Columbia River Plateau, Washington-Oregon; Mauna Loa, Kilauea, Hawaii

Alkali basalts: Ocean floors; Inner Hebrides, Scotland; Antrim, Northern Ireland; Iceland; Faroe Islands; Mauna Kea, Mauna Loa, Kilauea, Hawaii

Leucite basalts: Italy; Germany; East Africa; Montana; Wyoming; Arizona

Nepheline basalts: Libya; Turkey; New Mexico

Amygdaloidal basalt: This kind of basalt with its white amygdale spots is easy to identify. Other igneous rocks do have amygdales, but they are rarely as large as in basalt. The black microscopically grained groundmass confirms its identity.

GLASSY ROCKS

When lava (or magma) cools quickly, there is no time for crystals to form within the mix, so the result is a glassy rock. Glassy rocks not only look like glass, though darker and cloudier, but they also shatter like it when struck with a hammer, giving sharp fragments. The main glassy rocks are obsidian, perlite and pitchstone, which are all inter-related and merge into each other, largely according to their water content.

Obsidian

Like rhyolite, obsidian forms from the same magma that solidifies as granite deep underground. When rhyolite magma approaches the surface, the reduction in pressure means that some of its water is lost as steam. This de-watered rhyolite magma becomes very thick and viscous. Indeed, it becomes so thick that crystals do not get a chance to grow before the erupted lava is chilled and frozen solid. The result is a rock that is just like solid glass except slightly harder, and often jet black. Obsidian lava is so thick that it advances at a snail's pace, and outcrops are usually quite small. It typically forms near the end of the volcanic cycle and creates just a small plug, a thin coating on rhyolite lava flows or a lining for rhyolite sills and dykes. Yet occasionally there are large flows of obsidian, as at Glass Buttes in Oregon and Valles Caldera in New Mexico, where there are layers of obsidian a few hundred metres thick. Such a flow occurred just 1,300 years ago at the Newberry volcano in Oregon.

Obsidian shatters like glass into sharp, conchoidal (curved) fragments. It quickly dulls with exposure, but when freshly broken obsidian gleams like polished glass. It is typically jet black, coloured by titanium oxide minerals, but streaks and swirls can make broken surfaces look like colour-flowed marbles, as in brown- and black-streaked 'mahogany' obsidian and 'midnight lace' obsidian, with its contorted streaks formed as the cooling lava rolled over and over. Iron oxides turn obsidian reddish or brownish, while gas bubbles and microcrystals make 'golden sheen' obsidians that shimmer iridescently in sunlight. Because it fractures with a beautifully curved glass-sharp edge, obsidian can easily be fashioned to make knife and axe blades, even better than flint. It can also take a high polish. As a result it was highly prized among early cultures. The Ancient Egyptians, the Aztecs and Mayans all used obsidian knives and arrowheads. So too did Native Americans. Because obsidian absorbs water once it is broken, some of these ancient obsidian artefacts can be accurately dated by measuring just how much water their surface layers have absorbed. Obsidian is rarely older than 20 million years, because it starts to alter as soon as it forms. In a process known as 'devitrification', the glass absorbs moisture and begins to form crystals and go cloudy.

Identification: Jet black obsidian, which looks like a lump of solid glass, is hard to mistake for any other rock, especially if it fractures conchoidally. But it can contain phenocrysts of quartz and microscopic crystals of feldspar.

Conchoidal fracture

Snowflake obsidian: Snowflake obsidian has white snowflake patches of the mineral christobalite. Sometimes, 'snowflakes' can form through devitrification, as moisture alters silica in the obsidian.

Grain size: None, obsidian is glassy
Texture: Occasional small phenocrysts or microlites (tiny crystals)
Structure: Breaks conchoidally; occasional spherulites (tiny radiating clusters of needlelike crystals); flow banding with alternating glassy and devitrified layers. Contortion of flowbands in unset lava creates 'midnight lace' obsidian.
Colour: Usually jet black, but presence of iron oxides turns it reddish and brownish, and inclusion of tiny gas bubbles gives it a golden sheen. Dark banding and mottled grey, green and yellow. Microscopic feldspar crystals create 'rainbow obsidian'.
Composition: As for rhyolite: Silica (74% average), Alumina (13.5%), Calcium and sodium oxides (less than 5%), Iron and magnesium oxides (less than 3.5%). Obsidian always includes a certain amount of water, often in the form of minute bubbles of water vapour trapped in the glass. These bubbles are usually visible under a magnifying glass.
Minerals: Quartz; Potassium feldspar (sanidine) and plagioclase feldspar (oligoclase); Biotite mica
Phenocrysts: Quartz; Christobalite
Microlites: Feldspar
Formation: Lava flows, dykes and sills
Notable occurrences: Scotland; Eolie Island, Italy; Mount Hekla, Iceland; Mexico; Obsidian Cliff (Yellowstone), Wyoming; Arizona; Colorado; Valles Caldera, New Mexico; Big Obsidian Flow (Newberry), Glass Buttes, Oregon

Perlite

Like obsidian, perlite is a natural glass that forms from rhyolite lava, but rather than being shiny black, perlite is grey like dirty snow. Also while obsidian contains very little water, perlite gets its name because it contains concentric cracks that make the rock break into tiny pearl-like balls. Unlike obsidian, perlite contains water (2–5 per cent) because it cools so quickly that water has no time to escape. Once formed, it goes on absorbing water from its surroundings. Each little pearl is like a balloon full of water, and this is what makes perlite a rather amazing material. When heated to 871°C/1,600°F, the water evaporates, and the steam turns each pearl into a bubble, inflating the perlite like popcorn up to 20 times its original volume. This creates an incredibly light, gas-filled material, which is used for all kinds of insulation, for both heat and sound. Many roofing tiles contain perlite as does pipe insulation. It is also used instead of sand to make lightweight concrete. Horticulturalists often grow plants in a perlite mix instead of soil because of its good aeration and water retention.

Identification: Perlite looks a bit like dirty ice, and lumps of it can look rather like old snowballs. It has a glassy texture, with no crystalline structure, but is often dotted with phenocrysts, much more so than obsidian.

Perlite pebble

Grain size: None, perlite is glassy
Texture: Occasional small phenocrysts; when there are a lot of phenocrysts, the rock becomes 'vitrophyre'
Structure: Concentric cracks, which mean the rock breaks into pearl-shaped balls
Colour: Grey or greenish, but may be brown, blue or red
Composition: As for rhyolite: Silica (74% average), Alumina (13.5%), Calcium and sodium oxides (<5%), Iron and magnesium oxides (<3.5%)
Minerals: Quartz; Potassium feldspar (sanidine) and plagioclase feldspar (oligoclase); Biotite mica
Phenocrysts: Quartz, sanidine, oligoclase or, rarely, biotite or hornblende
Formation: Lava flows, dykes
Notable occurrences: Greece; Turkey; New Mexico; Sierra Nevada, California; Utah; Oregon

Apache tears

Sometimes obsidian alters to perlite when it absorbs water during or after cooling. The water is gradually absorbed along cracks caused by the cooling process, turning more and more of the obsidian to balls of perlite. As more water is absorbed, these spread out in concentric circles through the obsidian. Eventually, all that is left is a small core of obsidian embedded in a mass of perlite. As the process goes on, the perlite is broken by weathering, leaving just a few isolated nodules of obsidian, rounded by wind and water into natural marbles, called 'Apache tears'. They got their name from the stones at Apache Leap Mountain near Superior in Arizona. Legend has it that in the 1800s Apache warriors were trapped at the top of a cliff on this mountain by pursuing US cavalry. Rather than surrendering to their enemies, the apaches leaped to their death. The tears of their wives and children are said to have fallen to the ground here and the Great Spirit, looking down, turned them into the Apache tears so that the courage of the warriors might be remembered forever.

Pitchstone

Pitchstone is a glassy volcanic rock occurring famously in the Hebridean islands of Scotland, where it was used in the Stone Age to make blades. Some pitchstone is high in silica and forms from the same granite-like magma as rhyolite. Other pitchstones are lower in silica and more akin to trachyte or even andesite. Unlike obsidian, pitchstone contains a lot of water (up to 10 per cent) and is dull in lustre – especially older pitchstones that have become almost completely devitrified (lost their glassiness) and look pretty much like rhyolite. Many pitchstones contain phenocrysts arranged in wavy tracks reflecting the flow of the magma. Pitchstones are often mixed in with crystalline volcanic rocks, and may have formed when water driven out of the crystallizing rock was taken up by the glassy pitchstone.

Grain size: Glassy, or cryptocrystalline
Texture: Abundant phenocrysts; when phenocrysts dominate, the rock becomes 'vitrophyre' or pitchstone porphyry
Structure: Wavy flow streaks. Breaks to poorly defined conchoidal fracture
Colour: Streaked, mottled, or uniform black, brown, red, green
Composition: Quartz (variable), Potassium feldspar and plagioclase feldspar, Biotite mica
Phenocrysts: Quartz, potassium feldspar, plagioclase, or, rarely, pyroxene or hornblende
Formation: Dykes and sills
Notable occurrences: Arran, Eigg, Skye (Hebrides), Scotland; Chemnitz, Meissen, Germany; Lipari, Italy; Urals, Russia; Japan; New Zealand; Oregon; Colorado; Utah; California

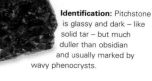

Identification: Pitchstone is glassy and dark – like solid tar – but much duller than obsidian and usually marked by wavy phenocrysts.

VOLCANIC FROTH AND ASH

Not all volcanic rocks are formed from molten lava. Pumices form from glassy lava so filled with gas
bubbles that it becomes a froth. Tuffs form from ash and pyroclasts – fragments of solid magma and rock
shattered by an explosive eruption. Some falls to the ground and only gradually consolidates into rock.
Some rushes out in flows so hot that material is literally welded together, creating solid ignimbrite.

Pumice

Floating rock: Pumice
will float for several
months before becoming
waterlogged and sinking.

Pumice is the only rock that floats.
It is solidified lumps of rhyolitic or
dacitic lava froth so full of holes
that it is less dense than water.
When the lava erupted, the release
of pressure made gases dissolved in it
effervesce – like unscrewing a shaken
fizzy drink bottle – and form bubbles
that blew the lava up into a froth.
Had it stayed under pressure it would have formed obsidian.
Basalt and andesite lavas form froth rocks, too, called
scoria, but because these lavas are so fluid, gases can escape.
So scoria contains fewer holes than pumice and won't float.
A basalt pumice does form in Hawaii, however – and it is
even lighter than rhyolite pumice, and black! As volcanoes
have erupted through time, tiny fragments of
pumice have been scattered all over the world,
and now coat the deep ocean floor. Some has
come from undersea eruptions, but much came
from fragments falling on the ocean after big
eruptions, then floating for months before
becoming waterlogged and sinking. Ground up,
pumice is the abrasive used to 'stonewash' jeans
among other things. Commercially the word pumice
refers only to large stones; grains are called pumicite.
Pozzolan is a pumicite mixed with lime to make cement.

Identification: Fresh pumice is
very easy to identify since it is
whitish, full of holes and so light
that it actually floats, until it
becomes waterlogged. However,
pumice retains this lightness for
only a short time geologically.
Soon enough all the holes are
infilled with secondary minerals
and it is no longer buoyant, and
the glassy solid material
becomes devitrified.

Pumice is often full
of air holes like this

Grain size: None, it is glassy
Texture: Like a foam
Structure: The solid glass
forms threads and fibres
surrounding rounded or
elongated holes depending
on the flow of the lava.
Cavities may be infilled when
percolating water deposits
secondary minerals.
Colour: Usually white or light
grey; scoria is black or brown
Composition: The same as
for rhyolite: Silica (74%
average); Alumina (13.5%);
Calcium and sodium oxides
(<5%); Iron and magnesium
oxides (2%)
Minerals: Quartz; Potassium
feldspar (sanidine) and
plagioclase feldspar (oligoclase);
Biotite mica
Formation:
Lava flows
and pyroclasts
**Notable
occurrences**:
Pozzola, Italy;
Greece; Spain;
Turkey; Chile;
Arizona; California; New
Mexico; Oregon

Banded tuff

Identification: Banded tuff is
streaked with dark glass
and welded ash.

Tuff is rock formed from consolidated volcanic
ash. It typically shows layering as larger
heavier particles land first in each eruption
(unlike ignimbrite). But every now and then
it can show a much more distinctive
banded pattern. This is sometimes due to
patterns created when hot ash is
welded, and sometimes a result of
last-minute mixing of magma from
different sources. The two never
mix perfectly, and the result is that
when the pyroclasts finally settle,
they accumulate in layers reflecting the
different mixes of magma. Such banded tuffs were found after the cataclysmic
1912 eruption of Novarupta in the Valley of 10,000 Smokes in Katmai, Alaska.

Grain size: Welded into
glassy or amorphous mass
Texture: Said to be eutaxitic
when it contains flammes
(glassy pancakes of pumice)
Structure: Welding and
mixing creates other bands
Colour: Grey to black, may
be turned pink by weathering
Composition: Variable –
usually from rhyolite or
tachyte glass
Formation: Ashfalls and
pyroclastic flows
Notable occurrences:
Charnwood Forest (Leics),
England; Mato Grosso, Brazil;
Santa Cruz, California;
Cripple Creek, Colorado;
Katmai, Alaska

Vesuvius and Pompeii
Mount Vesuvius is one of the world's most famous volcanoes. There have been eight major eruptions so far, the most recent in 1906 and 1944. Vesuvius is a composite volcano typical of subduction zones, supplied by trachyte and andesite magma. Its explosive, Plinian-type eruptions send up towering columns of ash, pumice and bombs, which smother the surrounding area, or collapse to push out devastating pyroclastic flows. In the terrible eruption of AD79, witnessed by the Roman writer Pliny, pyroclastic surges incinerated the mountainside town of Herculaneum (original settlement shown in the foreground, above), while ashfall completely buried Pompeii, killing the inhabitants, but preserving their homes perfectly. The eruption was catastrophic, but Vesuvius is a small cone sitting in the caldera of Monte Somma, a giant volcano that erupted 35,000 years ago with a force that would make the Pompeii eruption seem a mere puff. Over 30,000km²/11,583 sq miles of the land around the Bay of Naples is Campanian ignimbrite, created in a vast pyroclastic flow in this eruption.

Ignimbrite

No rock is created in such a rapid and dramatic way as ignimbrite. It is the rock that forms when pyroclastic flows and surges finally come to rest. Its name is Latin for 'fire cloud', and is very apt. Pyroclastic flows are clouds of glowing ash, cinders and hot gases that roar down from an eruption at jet plane speeds and temperatures of 450°C/842°F or more. Many flows are still so hot when they come to a halt and settle that volcanic fragments within them are instantly welded together. Towards the base of flows, the heat can squeeze pumice fragments flat to create pancake shapes called *flamme* (flames).

Grain size: Varied, mostly less than 2mm/0.08in
Texture: Like a fruitcake. Said to be eutaxitic when it contains flammes (glassy pancakes of pumice).
Structure: Large pebbles of pumice in finer mass of glass fragments. Sometimes shows flowbanding and layering; welding creates other bands.
Colour: Grey to bluish grey, may be turned pink by weathering
Composition: Variable – usually from rhyolite or tachyte glass
Phenocrysts: Feldspar
Formation: Pyroclastic flows
Notable occurrences: Widespread; Mount Vesuvius, Italy; Hunter Valley (New South Wales), Australia; Coromandel, New Zealand; Ria Loa, Chile; Mount St Helens, Washington

Identification: Ignimbrite has a distinctive dark fruitcake look, with its isolated phenocrysts and flammes of pumice, but can easily be mistaken for lava flow rock.

Lithic tuff

Ash thrown out by volcanoes settles on the ground like falling snow, building up in drifts. At first, ash-fall is just loose dust. But in time, it packs down and becomes consolidated into a soft, porous solid rock called tuff. Sometimes lithification (turning to stone) is helped by the way glass in the ash is turned by the weather into clay and zeolite cement. Tuffs vary widely in texture and composition, and older tuffs have lost most of their original texture through recrystallization. They can be classified into three kinds according to the predominant fragments in the ash: 'lithic' tuffs, made mostly of chunks of broken rock; 'vitric' tuff, made mostly of shards of volcanic glass; and 'crystal' tuff, made mostly of small crystals such as feldspar, augite and hornblende. Most ash grains in tuff are less than 2mm/0.08in across, but tuff can also contain pebble-sized fragments called lapilli. Wind can scatter ash over huge distances, but lapilli land close to the volcano. Those falling close enough may be still hot enough to weld ash together, forming welded tuff.

Identification: Tuff is much softer than any other volcanic rock, easy to scratch with a knife. Although there may be visible crystals in new tuffs, they are very unevenly distributed and rather shapeless.

Grain size: Varied, mostly less than 2mm/0.08in
Texture: Like a dense sponge cake
Structure: Tuffs are usually layered, rather like sedimentary rocks with the heaviest particles that drop first in each ashfall at the base of each layer
Colour: Grey, black, very variable – older basaltic tuffs may be turned green as original minerals are altered to chlorite
Composition: Variable
Minerals: Variable
Formation: Ashfall and pyroclasts
Notable occurrences: Widespread, including Santorini, Greece

VOLCANIC DEBRIS

When explosive volcanoes erupt, 90 per cent of the solid material ejected is not lava but pyroclastic material. The word 'pyroclastic' means 'fire-broken'. Pyroclasts are fragments of old magma, fresh magma and basement rock shattered into ash, lapilli (stones) and bombs (boulders) – collectively called tephra – by the explosive force of the eruption and scattered far and wide around the volcano.

Lapilli

Lapilli are pyroclasts usually thrown out by explosive volcanic eruptions. Lapilli is Italian for 'little stones' and geologists define them as pyroclasts 4–64mm/0.17–2.5in across – in other words, between the size of a pea and that of a walnut. Anything larger is classified as a bomb; anything smaller is ash. Some lapilli are globules of fresh, liquid magma. Some are fragments of exploded magma. Some are fragments chipped off basement rock by the eruption. Globules of magma can sometimes cool and solidify into a teardrop shape as they fly through the air. Lapilli like these are called 'Pelée's tears', after the Hawaiian goddess of volcanoes. Pelée's tears often trail a thread of liquid lava that chills in midair into a golden brown hair-like filament called Pelée's hair. Occasionally, nut-sized pellets called accretionary lapilli build up like hailstones in a thundercloud, as layers of ash cling to a drop of water. Froth in felsic lavas like rhyolite makes pumice lapilli that bob on water for months. Basaltic lava froth makes heavier scoria cinders, but occasionally forms reticulites. Reticulites are 98 per cent air bubbles – even lighter than pumice but so fragile they sink as the bubbles break and take in water.

Identification: Lapilli are little, light, cinder-like stones, often found in layers of ash or scattered around the foot of volcanoes. They may be glassy like these.

Size: 4–64mm/0.17–2.5in
Texture: Usually glassy and vesicular – that is, containing gas bubbles. Silicic magmas typically produce pumiceous (pumice-like) lapilli. Basalt typically produces scoria lapilli or, occasionally, light air-filled reticulite.
Colour: Black, grey or brown
Composition: Varies according to parent lava
Formation: Pyroclasts from fresh magma
Notable occurrences:
Cumbria, England; Stromboli, Mount Etna, Mount Vesuvius, Mount Vulture, Italy; Santorini, Greece; Toba, Tambora, Krakatoa, Indonesia; Citlaltépetl, Popocatépetl, Mexico; Kilauea Volcano, Mauna Loa, Hawaii; Yellowstone, Wyoming; Alamo, Texas

Tephra

This name was coined by Icelandic vulcanologist Sigurdur Thorarinsson in the 1950s. It comes from the Ancient Greek for 'ash', but is a general word used to describe all pyroclastic material thrown into the air by a volcano, including ash, lapilli and bombs of all kinds. The term is used only for material that falls from the air, but excludes pyroclastic flow material. Unlike tuff, tephra is loose material and turns into solid rock tuff only when cemented together. It varies in composition from scoria-fall (cinder) deposits to pumice-fall deposits. Scoria-falls are erupted in Strombolian-type eruptions and consist of basaltic to andesitic pyroclasts falling fairly close to the volcanic vent. These Strombolian scoria falls are typically dark in colour. Pumice-falls are blasted out in Plinian-type eruptions and consist of dacitic to rhyolitic pyroclasts often scattered over vast areas. Plinian-type pumice-falls are typically light in colour.

Identification: Tephra is a blanket term to describe any fragment thrown out by a volcano. It can be cinders like these, or ash and pumice, glass beads, and reticulite, to huge bombs and blocks.

Size: Full range of sizes
Texture: Usually glassy and vesicular – that is, containing gas bubbles
Structure: Mixture of ash, lapilli and bombs, usually sorted into layers with largest particles at the base and finest at the top
Colour: Black, brown or grey
Composition: Varies according to parent lava
Formation: From falling pyroclasts
Notable occurrences:
Many locations including: Surtsey, Hekla, Iceland; Stromboli, Mount Etna, Mount Vesuvius, Italy; Santorini, Greece; Toba, Tambora, Krakatoa, Indonesia; Citlaltépetl, Popocatépetl, Mexico; Yellowstone, Wyoming

Blocks and bombs

Blocks and bombs are large fragments of pyroclastic material. Blocks are chunks of broken magma. Bombs are large blobs of molten magma. Big eruptions can hurl blocks heavier than a truck 1km/0.6 miles from the vent. They can fling smaller bombs 20km/12 miles or even 80km/50 miles – at speeds approaching 75–200m/s (250–650ft/s), faster than a bullet. Most land close to the volcano, though. Bombs and blocks are not as heavy as they look, because they are usually full of holes. Volcanic bombs are fluid as they fly through the air and take on various, quite diverse shapes, according to just how fluid they are. Some end up as long, flat ribbon bombs. Some are so fluid that they are streamlined by force of motion as they fly into spindle bombs. Viscous lava bombs solidify at the surface to create breadcrust bombs as gas bubbles in the liquid interior expand and crack the surface like (very) crusty bread. 'Cow-dung' bombs hit the ground still molten and spread out in pancakes. Bombs can form various kinds of rock when they land, including tuff breccia, made from 25–75 per cent bombs, pyroclastic breccia (over 75 per cent bombs and blocks), agglomerates (over 75 per cent bombs) and agglutinate (made from spatters of basaltic lava still molten when it hits the ground).

Breadcrust bomb: Breadcrust bombs are made from blobs of viscous lava that crack on the surface to look like crusty bread. They earn their bomb name by occasionally exploding in mid-air as internal gas bubbles expand.

Grain size: Greater than 64mm/2.5in
Texture: Usually glassy and vesicular – that is, containing gas bubbles
Structure: Varies according to the way it cools in the air or on the ground, from the gas expansion cracks on the surface of breadcrust bombs to the long hair-like trails of Pelée's hair
Colour: Black or brown
Composition: Varies according to parent lava
Formation: Pyroclasts from fresh magma form into tuff breccia, pyroclastic breccia, agglomerate and agglutinate
Notable occurrences: Stromboli, Mount Etna, Italy; Kluchevskoy, Tolbachinksky (Kamchatka), Russia; Mauna Loa, Hawaii; Vanuatu; Mount Lassen, California; Yellowstone, Wyoming; Craters of the Moon, Idaho; Red Bomb Crater, Oregon

Perilous volcanic snow
Glowing streams of molten lava can be awe-inspiring, but tephra is also very dangerous. Ash quickly chokes people to death and buries vast areas under deep deposits. Rooftops covered in ash may collapse, crushing anyone beneath. Ash can also be a hazard to aircraft, as seen in 1982 when Java's Gallungang erupted, nearly bringing down two jet airliners as ash clogged their engines. Ash has also caused famine as it destroys vegetation. In 1815, Indonesia's Tambora sent ash up to 1,300km/808 miles away, and 80,000 people died of famine. The distance that ash travels depends on the height of the eruption column, the air temperature, and the wind direction and strength. The Krakatoa eruption of 1883 (Indonesia, above) spread 800,000km²/308,882 sq miles of ash, and burned the clothes of people 80km/50 miles away. The Toba (Indonesia) eruption of 75,000 years ago deposited 10cm/4in of ash 3,000km/1,864 miles away in India!

Block lava

Basalt lavas produce very fluid pahoehoe lava, which slurps along to freeze in rope-like coils, or more brittle, chunky *aa* lava. But more silicic lavas such as andesite and dacite produce a 'blocky' lava. Blocky lava is usually linked to stratovolcanoes with alternating layers of lava and pyroclasts. Lava flows from these volcanoes are very slow indeed, and the flow surface quickly congeals into a large rubble-like mass of blocks on top and at the front of the flow. The flow often follows narrow tongues called 'coulées' down the volcano, channelled by embankments or 'levées' of lava blocks either side. Rhyolite lavas are chunkier still and often develop dam-like ridges of blocks called ogives on the surface.

Pahoehoe lava: This fluid lava solidifies in rope-like coils from the hottest, most fluid basalt lava.

Texture: Chunky, angular blocks
Structure: Biggest, most angular chunks occur on the edges of the flow
Colour: Black, brown or grey
Composition: When andesitic: Silica (59% average), Iron and magnesium oxides (7.5%)
Formation: Andesite, dacite and rhyolite lava flows

Block lava: The surface of more viscous, silicic lava like andesite and dacite breaks into block-like chunks.

ULTRAMAFIC ROCKS

Ultramafic rocks are dark-coloured igneous rocks with even less silica than basalt, a lot of magnesium and iron, and made mostly of dark green or black olivines and pyroxenes. They form deep in the Earth's mantle and are usually brought to the surface in small quantities, often in masses no smaller than a fist or as large as a house, swept up as lumps in other magmas, or as the entire crust is uplifted by tectonic movements.

Picrite

Identification: Picrite is best identified by its setting, its dark green to black colour, its slightly shiny, sugary look and its even texture – not mottled like peridotite nor with plagioclase feldspar laths like dolerite (diabase), which are both often found with picrite.

Picrites are dark, heavy rocks rather similar to peridotite. Both form deep in the Earth's mantle, and both are rich in dark green olivine and brown augite. But while peridotite is often carried up into large intrusive masses, along with gabbro, norite and pyroxenite, picrite tends to be found in sills and intrusive sheets. Although picrite magma forms only under extreme pressures deep in the mantle, it is the one ultramafic rock that normally erupts on the surface as lava, as it did in the 1959 eruption of Kilauea in Hawaii. In this eruption, gigantic fire fountains shot out picrite lavas containing as much as 30 per cent olivine. Yet for picrite to erupt as lava like this, temperatures must be very high indeed – which is why it is often linked to hotspot volcanoes such as those in Hawaii.

Picrite is also often found in association with basalt as part of the ocean floor, which is why many of the best known occurrences of picrite are in ophiolites – chunks of the ocean floor brought to the surface by massive tectonic movements.

Occasionally, picrite can occur in substantial quantities in flood basalts, as it does in India's Deccan and South Africa's Karoo, but most flood basalts have a fairly low picrite content.

Picrite is very rich in magnesium and iron. Indeed, one definition of picrite is that it has at least 18 per cent magnesium oxide by weight. Komatiite is similar but has even less sodium and potassium oxide. Picrite's iron content can be so high that it is actually slightly magnetic. Some picrites, though, are especially rich in hornblende (see right). Others are especially rich in augite, like a few of those of Devon and Cornwall in England, which are sometimes called palaeopicrites because they formed well over half a billion years ago in the Palaeozoic era. Most picrites in this part of the world date more specifically to the Devonian period, 408–360 million years ago.

Grain size: Moderately fine-grained (salt-sized)
Texture: Granular
Structure: Evenly textured
Colour: Dark green to black
Composition: Silica (47% average), Alumina (10%), Calcium and sodium oxides (10%), Iron and magnesium oxides (31%)
Minerals: Olivine; Clinopyroxene (augite); Orthopyroxene (enstatite); Biotite mica; Hornblende; Plagioclase feldspar
Accessories: Apatite, melilite, sphene, biotite, spinel, hornblende
Phenocrysts: Green olivine or red brown augite
Formation: In extrusive lavas at mid-ocean ridges and hotspots, dykes and sills. Appears in ophiolites and flood basalt lava plateaux.
Notable occurrences: Rhum (Hebrides), Inchcolm Island, Midland Valley, Scotland; Devon, Cornwall, Sark (Channel Islands), England; Wicklow, Ireland; Nassau, Fichtelberg, Germany; Troodos, Cyprus; Gran Canaria; Oman; Madagascar; Karoo, South Africa; Deccan, India; Dongwhazi, Kunlun, China; Tasmania; Hawaii; Yellowstone, Wyoming; Klamath Mts, California; Oregon; Hudson River; Alabama; Montana

Hornblende picrite: The minerals in picrite often decompose quite quickly. Olivine is replaced by green, yellow and red fibres of serpentine, while augite is replaced by chlorite or hornblende. Only hornblende remains unaltered, creating hornblende picrite. Many ancient picrites are like this, such as those found in Gwynedd and Anglesey in Wales and on Sark in the English Channel Islands.

Pyroxenite

Like picrite and peridotite, pyroxenite is an ultramafic rock formed from magmas that develop deep in the Earth's mantle. What makes pyroxenite different from these rocks – though similar to wehrlite and clinopyroxenite – is that it contains a high proportion of clinopyroxene (usually the mineral augite, clinopyroxenite) or orthopyroxene (enstatite or hypersthene, orthopyroxenite), at the expense of olivine. Sometimes, pyroxenites are inclusions in other magmas, and look rather like obsidian: shiny black and fracturing into the same sharp, conchoidal (curved) fragments. Pyroxenites rarely occur alone. Very often, they form layered complexes with other plutonic (deep-forming) igneous rocks such as gabbro and norite. In the Bushveld complex of South Africa, for instance, gabbro, norite and pyroxenite layers are interwoven, with more pyroxenite layers at the base and more gabbro layers at the top. Occasionally pyroxenite-like rocks may form when certain limestones are altered by contact with hot magma, but these are more properly called pyroxene hornfels.

Identification: Pyroxenites are very much plutonic, which means they are almost entirely coarse-grained, often containing individual crystals several centimetres long. They are hard to distinguish from similar dunites, hornblendites, and melilitites, except by laboratory tests.

Grain size: Medium to coarse
Texture: Granular
Structure: May be layered
Colour: Green to black
Composition: Silica (47% average), Alumina (10%), Calcium and sodium oxides (10%), Iron and magnesium oxides (31%)
Minerals: Clinopyroxene (augite); Orthopyroxene (enstatite, bronzite, hypersthene); Olivine; Biotite mica; Chromite; Hornblende
Accessories: Plagioclase, chromite, spinel, garnet, iron oxides, rutile, scapolite
Phenocrysts: Green olivine or red brown augite, orthopyroxene
Formation: Small intrusions such as stocks and dykes and in bands in layered gabbro
Notable occurrences: Shetland Islands, Scotland; Saxony, Germany; Bushveld complex, S Africa; New Zealand; Cortlandt (Hudson River), North Carolina

Ophiolites

The problem with learning about the oceanic crust is that it is so inaccessible. But in places segments of the crust have been heaved up by tectonic movements and incorporated into mountains. These segments are called ophiolites, and although rarely complete they afford a rich opportunity to study the ocean crust and its material. They are quite literally a slice through the ocean crust, and always tend to have the same layers. At the top is a blanket of sediment less than 1km/0.6 mile thick, composed of clay grains and dead plankton. Beneath this is a layer of basalt pillow lavas, sitting on sheets of gabbro dykes. Further down are more gabbros along with norite and olivine-rich gabbro. Further down still (3–4.75km/2–3 miles) are layers of chromite-rich dunite, wehrlite, peridotite and pyroxenite. At the base is an unlayered agglomeration of serpentinized peridotite with dunite, harzburgite and olivine pyroxenite. There are famous ophiolites at Troodos in Cyprus. Others are at Josephine County in Oregon.

Melilitite and nephelinite

These deep-forming ultramafic rocks are often linked to hot spots and rifts, where material from deep down is brought to the surface. They are typically 'coughed up' from the mantle as xenoliths, masses in other magmas, and probably form when mafic magmas mix with peridotite. Although they are rare, they are found all over the world. They are both highly alkaline, like carbonatite, lamprophyre and kimberlite. Nephelinite is essentially nepheline and clinopyroxene, plus olivine and iron and titanium oxides. Melilitite is basically the same and there is a gradation between the two, but melilitite has the mineral melilite instead of nepheline.

Grain size: Medium to coarse
Texture: Granular
Structure: May be layered
Colour: Green, dark green to black
Composition: Silica (less than 42% average), Alumina (15%), Calcium and sodium oxides (17.5%), Iron and magnesium oxides (18.5%)
Minerals: Feldspathoid (nepheline or melilite); Clinopyroxene (augite); Olivine; Perovskite
Phenocrysts: Green olivine or red-brown augite
Formation: Both intrusive and extrusive, forming xenoliths, lava or pyroclasts at hot spot and rift volcanoes
Notable occurrences: Umbria, Italy; Rhine, Germany; East Greenland; Ol Doinyo Lengai, Tanzania; Western Cape, S Africa; Tasmania; Amazon; Sword Mount (Washington Co), Maryland

Identification: Melilitite is almost as evenly dark as basalt, but the grains are coarse enough to be visible. It is often found in kimberlites or with carbonatites.

OLIVINE-RICH ROCKS and CARBONATITE

Some igneous rocks formed deep in the mantle are especially rich in the green mineral olivine, which is one of the most common minerals on Earth, though most of it is in the mantle rather than in the crust. These rocks are brought up as xenoliths in other magmas, or form in lower parts of the sea floor or near the surface in small quantities as they separate from hot magmas, like those over hotspots and at rifts.

Peridotite

Peridotite is one of the main materials of the upper mantle, and both basaltic and gabbroic magma are thought to be produced by the melting of peridotite in the mantle. So ultimately, mantle peridotite is the source of most of the material that makes up the rocks of the Earth's crust. Laboratory tests have shown that if peridotite is heated to melting point, the melt is basalt. Partial melting seems to give picrites and komatiites.

Some peridotites appear low down in ophiolite complexes, layered along with pyroxenite. Others seem to accumulate at the bottom of gabbro intrusions when olivines crystallize and settle out. A few peridotites are squeezed upwards in deep volcanic pipes. Rare chunks of peridotite reach the surface as solid lumps, swept up as xenoliths in liquid magma – spinel peridotites in basalt, basanite, nephelinite and occasionally andesite, and garnet peridotites in kimberlites and lamproites.

Peridotite is the least siliceous of all the igneous rocks, with less than 46 per cent silica by weight, and contains almost no feldspar. So it is a very dark coloured rock, with small pale green crystals of olivine set in a mass of dark pyroxene and hornblende. There are actually several varieties of peridotite. Besides olivine, lherzolite contains both clinopyroxene (augite) and orthopyroxene (enstatite); wehrlite contains only clinopyroxene; harzburgite contains only orthopyroxene.

A special form of peridotite called kimberlite is, with related lamproite, the world's only primary source of diamonds (see Kimberlites and lamproites, right). But other peridotites are major sources of nickel and chromium minerals, platinum and talc, and chrysotile asbestos, too. In warm, humid places, peridotite has weathered to soils rich in iron, nickel, cobalt and chromium, which may one day be exploited as ores on a large scale.

Identification: Peridotite is a dark green to black, medium to coarse grained rock that looks rather like sugar stained with a dark green engine oil. Often it is studded with small pale green balls of olivine, or more rarely with red garnets.

Grain size: Medium to coarse (sugar-sized)
Texture: Granular. Frequently poikilitic, which means small round crystals (of pale green olivine) embedded in large irregular masses (of pyroxene and hornblende). Rarely porphyritic.
Structure: May be layered
Colour: Dull green to black
Composition: Silica (44–45.5%), Alumina (2–4%), Calcium and sodium oxides (4–6%), Iron and magnesium oxides (41–49%)
Minerals: *Wehrlite*: Olivine; Clinopyroxene (augite); *Lherzolite*: Olivine; Clinopyroxene (augite); Orthopyroxene (enstatite); *Harzburgite*: Olivine; Orthopyroxene (enstatite)
Minor and accessory minerals: Biotite mica, hornblende, chromite, garnet, picotite, corundum, platinum, awaruite
Formation: Interlayered with other ultramafic rocks in ophiolite complexes; settled out of gabbroic intrusions; in volcanic pipes, dykes and other small intrusions; and as xenoliths in basalt
Notable occurrences: Lizard (Cornwall), England; Anglesey, Wales; Skye, Ayrshire, Scotland; Harzburg, Odenwald, Silesia, Germany; Lherz (Pyrenees), France; Hungary; Norway; Finland; New Zealand; New York; Maryland

Garnet peridotite: Most peridotite has traces of garnet or spinel, but sometimes, especially in lherzolite, small crystals can form like cherries in a bun. Some are big enough to chip out as gems. Garnet and spinel peridotites, or rather lherzolites, are found as xenoliths in basaltic rocks and kimberlites, and are thought to form in the upper mantle.

Garnet

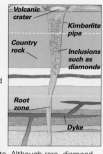

Volcanic crater · Kimberlite pipe · Country rock · Inclusions such as diamonds · Root zone · Dyke

Dunite

This rock is a kind of peridotite that is almost pure olivine. When freshly exposed it is olive green like olivine. But it gets its name from Mount Dun, the dun-coloured mountain in New Zealand, because dunite turns brown when weathered. Although some is formed in the mantle and appears in ophiolite complexes, much of it forms when olivine crystallizes out and sinks to the bottom of an intrusion of basic magma like gabbro. It is rich in a number of rare minerals, and is the world's major source of chromium ore.

Identification: Dunite is almost pure olivine, so when fresh it has a distinctive green colour as well as sugary texture. More usually though it is a tan brown, the colour it turns when exposed to the air.

Grain size: Medium to coarse (sugar-sized)
Texture: Granular
Structure: May be layered
Colour: Dull green, weathering to brown
Composition: Silica (41%), Alumina (2%), Calcium and sodium oxides (1%), Iron and magnesium oxides (54.5%)
Minerals: Olivine; Clinopyroxene (augite); Orthopyroxene (enstatite)
Minor minerals: Chromite, magnetite, ilmenite, pyrrhotite, pyroxene
Formation: At the base of gabbro intrusions or in the mantle
Notable occurrences: Lizard (Cornwall), England; Anglesey, Wales; Skye, Ayrshire, Scotland; Harzburg, Odenwald, Silesia, Germany; Lherz (Pyrenees), France; Hungary; Norway; Finland; New Zealand; New York; Maryland

Carbonatite

Almost all volcanic activity involves magmas based on silicates, so the existence of igneous rocks containing over 50 per cent carbonates is surprising. When first discovered in the early 20th century, these carbonate rocks, called carbonatites, were thought by geologists to be simply limestones moulded by the heat of silicate magmas, especially as many carbonatites occur interleaved with silicate rocks in volcanic structures called complexes. In fact, carbonatites originate as magmas in the mantle, just like other magmas. They are the coolest of all magmas, melting at just 540°C/1,004°F (compared to over 1,100°C/2,012°F for basalt), and carbonatite lavas look like liquid mud. Carbonatites are forcing geologists to rethink their understanding of mantle processes and magma production. There are 350 or so known carbonatite intrusions, over half of them in Africa. Most are quite small and occur beneath volcanoes, interweaving with formations of other magmas to create complexes. Even more surprisingly, there are a few places in the East African Rift Valley, and at Kaiserstuhl in the Rhine, where carbonatites have erupted on the surface. In 1960, it was realized that Oldoinyo Lengai in Tanzania is actually a carbonatite volcano.

Identification: Carbonatite often looks just like marble. It is the only white or even light grey igneous rock, and it is often possible to distinguish it from marble only by the place where it occurs and by complex laboratory tests.

Grain size: Medium to fine
Texture: Granular
Colour: White to grey (Sövite), cream/yellow (Beforsite), yellow-brown (Ferro-carbonatite)
Composition: Over 50% carbonates (<3% silicates)
Minerals: Calcite (Sövite carbonatite); Dolomite (Beforsite carbonatite) or Ankerite (Ferro-carbonatite). Plus apatite, phlogopite, aegerine, magnetite
Accessory minerals: Many
Trace elements: Includes barium, zircon, niobium, molybdenum, yttrium
Formation: Intrusions in complexes with ijolites, syenites, fenites; eruptions of lava and ash with nephelinites
Notable occurrences: *Intrusions*: Fen, Norway; Kola, Russia; Loolekop, S Africa; Okurusu, Namibia; Ambar Dongar, India; Bayan Obo, Mongolia; Jacupiranga, Brazil; Oka, Quebec; Mountain P, CA *Volcanoes*: Kaiserstuhl (Rhine), Germany; Kerimasi, Mosonik, Shombole, Oldoinyo Lengai, East African Rift Valley

SYENITES

Unlike granites and their cousins, the syenites contain little or no quartz, but neither do they have large amounts of mafic minerals like peridotite and gabbro. Together, the syenites and their cousins make up a group called the alkaline igneous rocks, many of which are rich in feldspathoids, or 'foids', such as nepheline. The small amounts of mafic minerals they do contain are often in beautiful blues or greens.

Syenite

Sodalite syenite: Like granites, syenites can be divided into those in which the feldspars are mainly potash (potassium) or soda (sodium). Sodalite syenites such as pulaskite and nordmarkite are often perthitic, which means the sodium feldspar, mainly albite, intergrows with potash feldspar and gradually replaces it. Sodalite syenite can often be identified by lavender-blue specks of the feldspathoid sodalite.

Cancrinite syenite: This contains more of the feldspathoid mineral cancrinite than other syenites.

The name syenite was originally coined by the famous Roman scholar Pliny to describe the beautiful granite-like rocks quarried by the Ancient Egyptians at Syene (Aswan) on the Nile. In fact, these rocks are granite, and not syenite. But in the 18th century, the great German mineralogist A G Werner applied the term to some similar-looking rocks he found near Dresden, and the name now applies to igneous rocks like these.

The syenites are plutonic rocks that form massive intrusions rather like granite and, unlike the 'foid' syenites, are similarly rich in potassium feldspars. But syenites contain very much less quartz than granite. In fact, as the quartz content goes up they merge into granites, and since granites and syenites often form in the same places, quartz-rich syenites are hard to distinguish from quartz-poor granites.

Syenites are the underground equivalent of trachyte and are rich in alkali feldspars such as microcline and orthoclase. The orthoclase is white or pink and forms half the rock. So syenites, like the other alkaline igneous rocks, are all very light in colour – unlike mafic rocks that are low in quartz, such as basalt and peridotite, which are all dark.

Other minerals within syenites can make them among the most colourful and beautiful of igneous rocks, often displaying a shimmering iridescence when cut and polished. This is why they are such popular ornamental stones. Sodalite gives a lavender-blue tinge to soda syenite, while green aegerine can make aegerine syenite green. Syenites can sometimes be distinguished from granites because they contain dark blue needles of amphibole, while granite contains black prism-shaped amphiboles or micas.

Syenites are divided into three main kinds, according to which dark, ferromagnesian (iron and magnesium) mineral predominates – augite, hornblende or biotite mica. The Larvik area of southern Norway is famous for its beautiful rainbow-sheened red or grey augite syenites called larvikites or 'blue pearl granite', which are often used for pillars and facades. Similar rocks are found in the Sawtooth Mountains of Texas.

Grain size: Phaneritic (coarse- to medium-grained). Can be pegmatitic.
Texture: Usually even grains, but can often be porphyritic
Structure: Often contains drusy cavities. Feldspars are can be perthitic – that is, they have intergrowths of potassium and sodium feldspar. May have veins of white albite.
Colour: Light red, pink, grey or white
Composition: Silica (59.5% average), Alumina (17%), Calcium and sodium oxides (11%), Iron and magnesium oxides (8%), Potassium oxides (5%).
Minerals: Potassium feldspar (microcline, orthoclase); Plagioclase feldspar (albite, oligoclase, andesine); Biotite mica; Amphibole (hornblende); Pyroxene (augite). Can contain up to 10% quartz before it becomes granite.
Accessories: Sphene, apatite, zircon, magnetite, pyrites, plus feldspathoids nepheline, sodalite, cancrinite, feklichevite and leucite. Larger quantities of these grade the rocks towards 'foid' syenites.
Phenocrysts: Potassium feldspar, diopside, plagioclase
Formation: Stocks, dykes and small intrusions, or interwoven with granites in large intrusions
Notable occurrences: Larvik, Norway; Saxony, Germany; Alps, Switzerland; Piedmont, Italy; Azores; Kovdor Massif (Kola Peninsula), Russia; Ilmaussaq, Greenland; Pilanesburg (Bushveld complex), South Africa; Alaska; Dharwar, India; Montregian Hills, Quebec; White Mountains, Vermont; Sawtooth Mountains, Texas; Arkansas; Montana

Nepheline or 'foid' syenites

Unlike ordinary syenites, the nepheline syenites contain no quartz whatsoever. Instead, they contain nepheline or another feldspathoid, or 'foid', such as sodalite or leucite – and the presence of quartz and these feldspathoids is mutually exclusive. Nepheline syenites are the intrusive equivalent of phonolite.

Foyaite: Grey, green or red nepheline syenite with a lot of microcline.

They are actually quite rare, and may form when soda-rich syenites and granites are altered in the melt, rather than forming distinct magmas. They often occur in ring dykes, and where granite magmas come into contact with limestones. There are many varieties of nepheline syenite, each with its own characteristics, including pink laurdalites from Laurdal in Norway, black mica-speckled miaskite from Miask in the Russian Urals, and nepheline-rich litchfieldite from Litchfield in Maine. But most come under the heading foyaites, after Foya in southern Portugal. The foyaites' odd chemistry means many rare minerals have been found in them, including eudialyte, eukolite, mosandrite, rinkite and lavenite.

Identification: Nepheline is often grey and can look like quartz, so it's easy to mistake nepheline syenite for granite or ordinary syenite. But nepheline syenite may have flow streaks and clots of dark minerals. And while quartz stays smooth on weathered surfaces, nepheline is pitted. The more nepheline there is, the greener it becomes. Blue sodalite and yellow cancrinite traces help to confirm the identity.

Grain size: Phaneritic (coarse). Can be pegmatitic.
Texture: Usually even grains, but often porphyritic
Structure: Flow streaks and clots of dark minerals
Colour: Usually grey, pink or yellow but can be greenish
Composition: Silica (42% av), Alumina (15%), Calcium & sodium oxides (17%), Iron & magnesium oxides (18.5%), Potassium oxides (5%)
Minerals: Potassium feldspar (orthoclase, microcline); Plagioclase (albite, oligoclase); Nepheline; Biotite; Hornblende; Pyroxene (augite, aegerine)
Accessories: Sphene, apatite, zircon, magnetite, pyrites, plus sodalite, cancrinite & leucite
Phenocrysts: Nepheline, albite
Formation: Stocks, ring dykes and small intrusions, with granites and syenites
Notable occurrences: Fen, Laurdal, Norway; Alnd I, Sweden; Foya, Portugal; Turkmenistan; Ilomba, Ulundi, Junguni, Malawi; Tasmania; Sierra de Tingua, Brazil; Quebec; White Mts, VT; Beemerville, NJ; Magnet C, AR

Monzonite

Monzonite gets its name from Monzoni in the Italian Tyrol. Like syenite, it is a plutonic rock that contains some, but not a lot of, quartz, and lies between the foids and the granites. In fact, it is actually what some geologists describe as the 'average' igneous rock, containing equal amounts of potassium and plagioclase feldspar, and lying halfway between the most acid and the most basic rocks in composition. Although by no means a rare rock, it only occurs in small masses, associated with and perhaps blending into gabbro and, at the margins of an intrusion, pyroxenite. There are three kinds of monzonite, depending on the presence of quartz, nepheline or olivine. Quartz monzonite is found in many mountain belts, and because it is such a tough rock often forms dramatic landforms.

Identification: Monzonite is light coloured like granite but contains less quartz.

Ancient Egyptian masters of stone
No ancient culture used stone with such skill as the Ancient Egyptians. The land is blessed with many superb stones for construction and carving, including monzonite, black and red granite, quartzite, limestone, sandstone and graywacke (siltstone). Red granite quarried at Syeneh Aswan (pictured above) was cut in single giant blocks for obelisks. Quartzite quarried at Gebelein was used for the famous Colossi of Memnon. Limestone from Tura, Beni Hassan and Ma'asa was used for the first pyramid, Djoser's. The Egyptians had a knowledge of geology and geological strata well beyond any culture of the time – and unmatched until recently. They were also amazingly skilled at cutting and shaping the stones. No-one knows quite how they did, since, strangely, stone-working does not appear in hieroglyphics. Some think they cut stones with copper saws, or bronze wire held in a bow. Others think they used emery stone.

Grain size: Medium
Texture: Irregular plates of orthoclase embedded in plagioclase
Structure: Crystals often have zones of different chemical composition
Colour: Dark grey
Composition: Silica (58.5% av), Alumina (17%), Calcium & sodium oxides (11%), Iron & magnesium oxides (9%), Potassium oxides (3%)
Minerals: Potassium feldspar (orthoclase); Plagioclase feldspar (labradorite, oligoclase); Pyroxene (augite, hypersthene, bronzite); Hornblende; Quartz, Nepheline or olivine; Biotite
Accessories: Apatite, zircon, magnetite, pyrites
Phenocrysts: Apatite, augite, orthoclase
Formation: Stocks, dykes and intrusions
Notable occurrences: Kentallen (Argyll), Scotland; Norway; Monzoni, Italy; Sakhalin Is, Russia; Yogo Peak (yogoite), Beaver Crk, Montana; Black Canyon, Colorado (quartz monzonite)

PLUTONIC ROCKS: GRANITE

Granites are by far the most common of all the plutonic rocks, rocks that form deep underground as giant batholiths, tens or even thousands of kilometres across. Although granites form only deep underground, they are often seen on the surface because their high quartz content makes them so tough they may survive long after softer rock around them has been worn away by weathering.

Granite

Granite lasts. Long after other rocks have crumbled away, intrusions of granite stand proud, like islands above the sea. The pointing finger of Rio's Sugar Loaf mountain, the sheer cliffs of Yosemite's El Capitan and the wild wastes of England's Dartmoor all stand testament to granite's ability to endure. For all these eminences are batholiths – huge intrusions of magma that formed entirely underground, and have simply been exposed through erosion of overlying rocks. The moors of Cornwall and Devon are all just knobs on a single large batholith that will, in 100 million years' time, be thoroughly denuded just as the moors are now.

Granite batholith complexes can be huge. The Patagonian batholith underlying the southern Andes is 1,900km/935 miles long and up to 65km/40 miles wide. That underlying the Sierra Nevada is almost as vast. In fact, there are giant granite batholiths underlying most of the world's great mountain ranges, both ancient, such as the Appalachians of North America, and recent, such as the Himalayas. Indeed, granite is always closely linked to mountain building, and the margins of continents where subduction is going on.

Although the bulk of granite is in batholiths, it can also form dykes and sills, and veins and intrusions of one granite can cut across another. Granite intrusions frequently interweave with country rocks, changing them where they come into contact. In many places, granite blends imperceptibly into metamorphic granite gneisses. Granite intrusions engulf lumps of country rock, which become xenoliths, often partly altering them on the surface at least.

Granite is a light-coloured, speckled rock and is very 'acidic' – that is, it has a high silica content (at least 70 per cent), and a high proportion of quartz (at least 20 per cent). It is essentially a mix of white or pink feldspar, pale quartz and small specks of black muscovite mica. Because it forms by cooling slowly deep underground, crystals in granite are almost invariably large enough to be visible to the naked eye. Most are at least a few millimetres. The largest white feldspar crystals can be prisms up to 20cm/8in long.

White granite: Granites all have soft black flakes of mica, the first to crumble from weathered surfaces. They all have pale crystals of quartz, too, and this never crumbles. It is the large feldspar crystals that typically give the colour varieties.

Feldspar

Pink granite: Most granites are light grey or pinkish in colour, but can also be dark grey or red. They are light grey if they are dominated by a white alkali feldspar, but pink or red if their alkali feldspar is pink or red. If there are both white and pink feldspars, the pink one is likely to be alkali feldspar and the white one is likely to be plagioclase.

Grain size: Phaneritic (coarse-grained); often pegmatitic
Texture: Normally granular, with large visible crystals and often porphyritic with large phenocrysts
Structure: Typically uniform, but may be banded. Xenoliths are common. Near the crown of batholiths, may be cracked into massive rectangular blocks.
Colour: Usually light coloured – mottled white, grey, pink or red, with black specks
Chemical composition: Silica (72% average), Alumina (14.5%), Calcium and sodium oxides (4.5%), Iron and magnesium oxides (less than 3.5%)
Minerals: Quartz; Potassium feldspar (microcline) and plagioclase feldspar (oligoclase); Mica; Hornblende
Accessories: Aegerine, zircon, apatite, magnetite, amphibole or pyroxene
Phenocrysts: Quartz, orthoclase and oligoclase feldspar, hornblende, biotite mica, augite
Formation: Intrusive: batholiths, stocks, bosses, sills, dykes
Notable occurrences: Syenogranite: Donegal, Ireland; Cairngorms, Scotland; northern Nigeria; Réunion Islands; Azores Islands; Canary Islands; Sugar Loaf Mountain (Rio de Janeiro), Brazil; Appalachian Mountains. Monzogranite: Cornwall, Devon, Lake District, England; Baltic shield, Finland and Sweden; Massif Central, France; Spain; Tatra Mts, Slovakia; Barrens, Newfoundland. Granodiorite: Andes; Rocky Mountains. Leucogranites: Himalayan Mountains.

The granite problem

Just how the world's giant granite batholiths formed has been the subject of fierce debate since the 18th century. The great Austrian geologist Eduard Suess could see they formed after the surrounding country rocks but, if so, how was such an enormous volume of granite accommodated? This 'room' problem has been at the heart of the controversy ever since.

In the mid-20th century, thinking developed into two opposing camps – the granitizers and the magmatists. The granitizers, like Doris Reynolds, believed that granite was formed when existing country rock is granitized (changed to granite) by 'metasomatic' processes. It all involved gases or fluids which came to be called 'ichor' after the lifeblood of the Greek gods. The idea was that country rock was metasomatized (altered) to granite by ichor emanating through it. That way, there was no 'room' problem. Magmatists, such as Norman Bowen, on the other hand, believed that granite formed from magma, not altered country rock. The magma was created at depth by the partial melting or 'anatexis' of the country rocks above.

Gradually, field observation of rock formations and laboratory experiments with melting rocks came down firmly on the side of the magmatists, and it is now widely accepted that most granites do indeed form by partial melting.

The theory is that it all takes place in the later stages of mountain building, as two tectonic plates crunch together. The collision not only buckles the plate edges to throw up mountain ranges, but also eventually creates such extreme pressures and temperatures that huge volumes of the mountain roots partially melt to create granite magma.

These hot granite melts are lighter than the rocks above, and so well up into them, like a blob of hot oil in a lava lamp, creating their own space. As they ascend they begin to cool and solidify, eventually turning to solid rock again.

Although the magmatists appear to have won this particular round of The Granite Problem, the controversy is by no means over. If granites do form by partial melting, for instance, then just which rocks have melted to form granite? Moreover, even among those who believed granite was a

magma, there were always those who believed that it formed not directly by melting of other rocks, but from basalt magmas changed gradually to granite as some chemicals crystallized and others didn't, a process known as fractional crystallization.

The evidence is that most continental granites – the great batholiths under mountains – do form by partial melting. But fractional crystallization may play a part in granites such as tonalites, which form in oceanic locations such as island arcs.

Granite varieties

Granite and granite-like rocks, called granitoids, come in so many subtly different varieties they have proved a nightmare to classify. One method is according to the balance of Quartz, Alkali Feldspar and Plagioclase feldspar (QAP) in their make-up. In the middle lies monzogranite in which alkali and plagioclase feldspar are fairly even. Syenogranite has more alkali feldspar, granodiorite has less. The extremes are alkali-rich alkali feldspar granite and plagioclase-rich tonalite. After the discovery that magmas making granites in eastern Australia came from both sedimentary and igneous rocks, and contained xenoliths of either type, geologists developed an 'alphabetic' way of dividing granites. I-type granites, rich in biotite mica and maybe hornblende, have a chemical make-up that implies they were formed from mafic igneous rocks. S-types, rich in both biotite and muscovite mica (plus garnet, cordierite and sillimanite), have a make-up that implies a sedimentary origin. M-types have a make-up implying they came from Earth's mantle. A-types have a make-up implying they are anorogenic – that is, they formed not in areas of mountain building but near hot spots and rifts.

Granophyre or porphyric microgranite: At the edges of intrusions, in thin intrusions and pegmatite, granite grains can be small. These microgranites often have phenocrysts of feldspar formed earlier and are related to graphic granites.

Quartz
Feldspar

Graphic granite: Graphic granite is an extraordinary kind of granite that occurs in pegmatites. The entire rock is one large crystal of pale feldspar, typically microcline. Embedded in the feldspar are thin wedge-shaped growths of quartz, giving an effect that looks like the cuneiform writing of Ancient Sumeria. It is thought that both the feldspar and the quartz formed at the same time.

GRANITE VARIETIES

Alkali feldspar granite: Granite with a very high proportion of alkali feldspar and almost no plagioclase

Biotite granite: Granite containing up to 20% biotite mica

Flaser granite: Granite-like gneiss, which has flattened feldspar crystals giving it a foliated look

Augite granite: Granite rich in dark augite

Tourmaline granite: Granite rich in black tourmaline

Leucogranite: Light-coloured granite with less than 30% of the norm for mafic minerals

Melanogranite: Dark-coloured granite with more than 30% of the norm for mafic minerals

Granite gneiss: Gneiss formed from a sedimentary or metamorphic rock and with the same mineral composition as granite

Peralkaline granite: Granite rich in alkaline feldspars and also alkaline amphiboles and pyroxenes such as aegerine and riebeckite

GRANITOIDS

There are many varieties of granite and granitoids (granite-like rocks). Some form only small patches within other granite outcrops, such as rapakivi granite and orbicular granite, with their highly distinctive round markings. Others, such as granodiorite and tonalite, have their own particular chemical and mineral make-ups and are individual rock types in their own right.

Rapakivi granites

Rapakivi granites get their name from the Finnish for 'rotten rock' because they weather easily. They are granites – usually syenogranites – with an unusual grain pattern called rapakivi texture. In rapakivi texture, large, oval crystals of alkali feldspar such as sanidine are encased in a plagioclase feldspar such as albite. The alkali feldspar forms first, then is enveloped by a 'reaction rim' of plagioclase as it reacts with the surrounding magma. One theory is that alkali feldspars from rhyolite magma react with plagioclase from basaltic magma as the two magmas mix. Another is that the effect occurs as pressure drops in a rising magma. Some geologists believe that the conditions for rapakivis occurred along continental rifts that never quite developed.
Rapakivi granites dating back 1,100 to 1,800 million years are found in a belt stretching right across Finland, Sweden and the Baltic to Labrador and the American south-west. They also occur in Brazil and Venezuela, and small pockets of more recent rapakivis are widely scattered.

Identification: Rapakivi granite is distinctive, with its pale rounds of feldspar embedded in a dark groundmass of mica, hornblende and quartz. When cut and polished, as here, rapakivi makes a popular decorative stone.

Grain size: Mixed
Texture: Oval feldspar phenocrysts up to 2cm/0.8in across embedded in a groundmass of small crystals
Structure: Typically uniform
Colour: Pink or tan K feldspar crystals rimmed with white albite in a dark groundmass
Chemical composition: Silica (72% average), Alumina (14.5%), Calcium and sodium oxides (4.5%), Iron and magnesium oxides (3.5%)
Minerals: Quartz; Potassium feldspar (sanidine) and plagioclase feldspar (albite); Mica
Phenocrysts: Usually alkali feldspar, but can be quartz or plagioclase feldspar
Formation: In syenogranites
Notable occurrences: S Finland; SE Sweden; St Petersburg, Russia; Estonia; Poland; Brazil; Venezuela; Labrador; Ontario; Maine; US mid-west and south-west

Orbicular granite

Occasionally, granites may contain small patches with an unusual texture called orbicular granite. This looks a little like rapakivi, but the balls, or 'orbicules', are bigger and they are not phenocrysts, but formations that develop around cores of foreign material in the magma. Each core may be a grain of another igneous rock (a small xenolith), but could also be a grain of granite. Alternating layers of pale feldspar and dark biotite or hornblende grow around the core, a process called 'rhythmic crystallization' in which first one mineral crystallizes then another as conditions change in the magma.

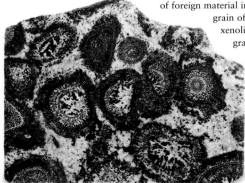

Identification: Orbicular granites are masses only a few metres across, and are easy to identify with their black and white rounds.

Grain size: Mixed
Texture: Large round orbicules 2–15cm/0.8–6in across embedded in a groundmass of small crystals
Structure: Typically uniform
Colour: Black and white layered phenocrysts in a light grey groundmass
Chemical composition: Silica (72% average), Alumina (14.5%), Calcium and sodium oxides (4.5%), Iron and magnesium oxides (3.5%)
Minerals: Quartz; Potassium feldspar (microcline) and plagioclase feldspar (oligoclase); Mica
Notable occurrences: Finland; Sweden; Waldviertel, Austria; Riesengebirge, Poland; Japan; New Zealand; Peru; Vermont

Granodiorite

Granodiorite is the intrusive equivalent of dacite and is the most abundant of all the granitoid rocks. It is very similar to granite but contains more plagioclase feldspar, and more mafic minerals (usually biotite and hornblende). In fact, granodiorite and granite often occur together, along with diorite, which contains even more plagioclase. In large batholiths, for instance, a single granodiorite magma may develop a granite heart and a skin of diorite or even tonalite, as minerals separate out in a particular way from the magma. This kind of process tends to happen only in large batholiths though. Where granite, granodiorite and diorite are all found in a smaller intrusion, the chances are they all came from separate magmas. Granitoid rocks such as granodiorite often occur in 'suites' – repeated associations of particular similar rocks. Granodiorite, for instance, is found in ancient, Archean formations along with tonalite and trondheimite. These TTG suites are among the world's oldest rocks, dating back more than two billion years, and are found all around the world, in places such as Lapland in Scandinavia, and the Big Horn Mountains of Wyoming.

Identification: Granodiorite is generally grey, and looks quite like granite, but it contains a higher proportion of dark minerals such as biotite mica and hornblende. Granite looks basically light grey with black specks. In granodiorite the grey and the black are more evenly balanced, giving a 'salt-and-pepper' look.

Grain size: Phaneritic (coarse)
Texture: Even texture; often porphyritic
Structure: Typically uniform
Colour: Black and white with pink potassium feldspar
Chemical composition: Silica (67%), Alumina (16%), Calcium & sodium oxides (7.5%), Iron & magnesium oxides (6%)
Minerals: Quartz; Plagioclase feldspar (oligoclase); Potassium feldspar (sanidine); Biotite mica; Hornblende
Accessories: Zircon, apatite, magnetite, ilmenite, sphene
Phenocrysts: Quartz or plagioclase feldspar
Formation: Intrusive: stocks, bosses, batholiths, sills, dykes
Notable occurrences: As for granite but also Aleutian Islands; Sonora, Mexico; Peninsular Mts, Baja California; Sierra Nevada, California.
TTG suites: Lapland, Finland; Barberton Mtn Land, S Africa; Pilbara, Yilgarn, Australia; Big Horn Mts, Wyoming

Tors

Granite-like rocks form underground, but are so tough they are often left standing proud after softer rocks are worn away. In some places, granite gives huge bare-rock cliffs. In others it gives rounded hills topped by outcrops of bare rock the size of a house, usually called by the ancient Cornish name 'tors'. Tors are a distinctive feature of the moors of England's south-west, but occur in many other places such as Scotland's Cairngorms and in South Africa, where they are called 'castle koppies'. There are several theories on how Cornish tors formed, but all are connected with the pattern of cracks, or 'joints', that develops parallel to the surface of the rock. Tors are tough clumps of rock that have survived after weaker surrounding granite was stripped away. One theory is that the softer granite was weathered in an earlier tropically warm age by natural chemicals seeping into the joints deep below ground. Another is that it was weathered by frost in the Ice Ages. The debris was, in both cases, probably swept away at the end of the Ice Ages by a process called solifluction, in which the water in frozen ground melts, turning the soil into a liquid mush that flows easily away.

Tonalite

This rock takes its name from Tonale in the Italian Alps near Monte Adamello. It is the quartz-rich, granitoid equivalent of diorite. It has the least potassium feldspar and the most plagioclase of any of the granite-like rocks, and also the most of the dark mafic minerals such as hornblende and biotite. The hornblende is often greenish rather than brown, while the biotite is pleochroic – that is, shows different colours from different directions. Tonalite is actually quite like granodiorite, and the two frequently occur together in TTG suites (see Granodiorite above). They have the same black and light grey 'salt-and-pepper' look, and can be hard to tell apart – tonalite has more black pepper.

Identification: Tonalites look very like granite, but are slightly darker and browner.

Grain size: Phaneritic (coarse)
Texture: Even texture; often porphyritic
Structure: Often threaded by veins of quartz and feldspar (aplites)
Colour: Pink or tan potassium feldspar crystals rimmed with white albite in a dark groundmass
Chemical composition: Silica (58%), Alumina (17%), Calcium and sodium oxides (10%), Iron and magnesium oxides (11%)
Minerals: Quartz; Plagioclase feldspar (oligoclase); Biotite mica; Hornblende
Accessories: Zircon, apatite, magnetite, orthite, sphene
Phenocrysts: Quartz or plagioclase feldspar
Formation: Intrusive: stocks, bosses, batholiths, sills, dykes
Notable occurrences: Galloway, Cairngorms, Scotland; Ireland; Rieserferner and Traversella (Tyrol), Austria/Italy; Andes, Patagonia; Sierra Nevada, California; Alaska.
TTG suites: Lapland, Finland; Barberton Mtn Land, S Africa; Pilbara, Yilgarn, Australia; Big Horn Mts, Wyoming.

GABBRO AND DIORITE

Gabbro and diorite occupy the middle ground of intrusive igneous rocks as far as the mineral balance goes. On the one hand, they contain only a little quartz and alkali feldspar – markedly less than the granitoid rocks. On the other, they contain only moderate amounts of olivine – much less than the ultramafics such as peridotite. Instead, they are made predominantly of plagioclase feldspar.

Diorite

Dark diorite:
Sometimes dark minerals can predominate over light in diorite.

Diorite is the coarse-grained, plutonic equivalent of andesite. It is darker than granitoids and contains heavier minerals, but it has a similar grain structure and forms in a similar way. Diorite is one of the igneous rocks that make their presence felt along continental margins where subduction of tectonic plates is throwing up mountain chains like the Andes. With granite, it forms the great long batholiths that underlie so many of these mountain chains. There is much less diorite than granite, and it often forms when rocks are caught up in a granite intrusion. Diorite contains much less quartz and alkali feldspar than granite, but more plagioclase feldspar than any other rock except anorthosite – over 75 per cent, even more than gabbro. The rest is mainly dark minerals such as hornblende and biotite. Diorite is similar to gabbro, but diorite's plagioclase tends to be oligoclase and andesine, whereas in gabbro much more of it is anorthite, bytownite and labradorite.

Identification: Diorite looks very similar to gabbro and it can be quite hard to tell them apart. Yet even though it contains very little quartz, its high plagioclase feldspar content means that light minerals are usually more prominent in diorite than gabbro. Essentially, diorite is light grey with patches of black, while gabbro is black with patches of light grey, but this is by no means a hard and fast rule.

Monzodiorite

This rock is midway between monzonite and diorite in composition. That means most monzonite contains some quartz, and a little more plagioclase than alkali feldspar, and fewer dark minerals than diorite. There is another kind of monzonite – feldspathoid monzodiorite – which contains nepheline or other foids instead of quartz. Monzodiorite used to be called syenodiorite, but IUGS (International Union of Geological Surveys) recommended it should be called monzodiorite to avoid confusion with monzonite and monzosyenite, which also lie in between syenite and diorite in composition, but contain more alkali feldspar.

Grain size: Phaneritic (coarse-grained), occasionally pegmatitic
Texture: Even-grained or porphyritic – often close together in the same rock
Structure: Foliation and xenoliths common
Colour: Speckled black and white, occasionally dark greenish or pinkish
Composition: Diorite: Silica (58.5% average), Alumina (17%), Calcium and sodium oxides (10.5%), Iron and magnesium oxides (11%), Potassium oxides (2%). Monzodiorite: Silica (58% average), Alumina (17%), Calcium and sodium oxides (11%), Iron and magnesium oxides (10%), Potassium oxides (2%).
Minerals: Plagioclase feldspar (oligoclase or andesine); Biotite mica; Amphibole (hornblende); Small amounts of pyroxene (augite), quartz and alkali feldspar (sanidine). Feldspathoid diorite and feldspathoid monzonite: Foids (nepheline) instead of quartz.
Accessories: Magnetite, apatite, zircon, titanite, olivine
Phenocrysts: Plagioclase feldspar, hornblende
Formation: Intrusive: sills, dykes, stocks, bosses, plus xenoliths in granite
Notable occurrences: Argyll, Scotland, Jersey (Channel Islands), England; Bavarian Forest, Black Forest, Harz, Odenwald, Germany; Finland; Washington; Massachusetts

Identification: Monzodiorite has the same 'salt-and-pepper' look as diorite and monzonite, but has slightly more 'pepper' than monzonite and slightly less than diorite.

Gabbro

Polished gabbro: Rarely, gabbro is cut and polished and used as a decorative stone.

Named after a town in Tuscany, Italy, by the great German geologist Christian Leopold von Buch, gabbro is the coarse-grained, intrusive equivalent of basalt and dolerite. It is a dark rock, basically made up of plagioclase and pyroxene. More pyroxene and less plagioclase merges it into peridotite; less pyroxene and more plagioclase blends it into diorite. Gabbro is a very widespread rock, especially in the oceanic crust, where it forms part of the ophiolite sequence. This is the sequence of rocks down through the ocean bed that develops either side of a mid-ocean rift. As pillow lavas and sheeted dykes of erupted basalt form the top of the sequence, so gabbro continually freezes in clumps from molten peridotite on to the magma chamber walls beneath as the walls move apart. Over millions of years, this has created a layer of gabbro underneath all the world's oceans.

Gabbro can also flow into sills and dykes, and occasionally, huge sheet complexes called lopoliths, like that at the Bushveld complex in South Africa, Duluth in Minnesota and Rhum in Scotland. In many places, the sinking of heavier minerals as they crystallized has created distinct layers in the gabbro with dark minerals concentrated at the bottom and light minerals at the top of each layer.

The Sudbury structure
In places around the world, there are huge, mostly ancient layered intrusions in which mafic magmas have crystallized in layers of different chemical composition. The biggest by far is the Bushveld complex in South Africa, which covers 65,000km²/25,000 sq miles. One of the most fascinating is at Sudbury in Ontario, Canada. The area is one of the world's richest sources of nickel copper, found in association with gabbro, and was once thought to be entirely igneous in origin. Now geologists have realized that it is not actually igneous at all but the huge impact crater of a meteorite that struck the ground here 1,850 million years ago – the biggest ever. This is clear from the shatter cones, rocks fractured in a cone shape by the impact (above). What makes this crater unusual, even for a meteor crater, is that it is oval – 200km/124 miles long and only about 100km/62 miles wide. Scientists believe, from mapping the geology of the area and creating various models, that it was created by a meteorite of some 10–19km (six to 12 miles) in width, which exploded on impact with the force of 10 billion Hiroshima bombs, fracturing the Earth's crust and bringing magma rich in mineral ores to the surface. This resulting heat melted the exposed granite and gneiss rocks into a glassy magma that deformed the crater. This mafic magma sits on top of the layered gabbro.

Identification: Gabbro can look similar to diorite, but tends to be darker as it contains the darker plagioclases (labradorite and bytownite) rather than the lighter plagioclases (oligoclase and andesine) of diorite. Gabbro is often ophitic (see texture, right), giving it a frosted look.

Gabbro is a very tough rock, which is why it is used widely for railway ballast and road metalling, but it is one of the least attractive of the intrusive igneous rocks, so is much less used as a decorative stone than the granites or syenites. However, it is virtually the only significant source of nickel, chromium and platinum minerals.

Grain size: Phaneritic (coarse-grained), occasionally pegmatitic
Texture: Even-grained or porphyritic – often close together in the same rock mass. Frequently ophitic, which means long light plagioclase feldspar crystals are enveloped by dark pyroxene (augite) crystals.
Structure: Foliation and xenoliths common. Often forms alternating layers, with mostly light minerals at the tops and mostly dark at the bottoms of the layers.
Colour: Black and white or grey, occasionally dark greenish or bluish
Composition: Silica (58% average), Alumina (17%), Calcium and sodium oxides (10.5%), Iron and magnesium oxides (11%), Potassium oxides (2%)
Minerals: Plagioclase feldspar (labradorite or bytownite); Pyroxene (augite); Olivine; Small amounts of amphibole (hornblende) and biotite mica; Very small amounts of quartz and alkali feldspar (sanidine)
Accessories: Magnetite, apatite, ilmenite, picotite, garnet
Phenocrysts: Plagioclase feldspar, hornblende
Formation: Intrusive: batholiths, lopoliths, sills, dykes, stocks, bosses, plus xenoliths in granite
Notable occurrences: Shetland, Skye, Rhum, Aberdeen, Argyll, Scotland; Pembroke, Wales; Lake District, Lizard (Cornwall), England; Skaergaard, Greenland; Bergen, Norway; Odenwald, Harz, Black Forest, Germany; Wallis, Switzerland; Bushveld complex, South Africa; Jimberlana, Windimurra, Western Australia; Eastern Canada; Baltimore, Maryland; Peekskill, New York; Stillwater, Montana; Duluth, Minnesota

Monzogabbro: Monzogabbro is gabbro with slightly less pyroxene and containing feldspathoids rather than quartz.

GABBROIC ROCKS

Gabbros are phaneritic (coarse-grained) rocks made of plagioclase feldspar, pyroxene and olivine. They are divided into various types according to how much of each they contain. Anorthosite is rich in plagioclase; troctolite is rich in plagioclase and olivine but not pyroxene; gabbro is rich in plagioclase and pyroxene but not olivine; essexite is rich in pyroxene. Norite is the midpoint.

Norite

Norite is a similar rock to gabbro, based on a mix of plagioclase, pyroxene and olivine, and the two often form in the same large, layered intrusions as the mix separates during crystallization. Norite contains very slightly less plagioclase than gabbro, but the real difference is that gabbro's pyroxene is a clinopyroxene such as augite, while norite's is an orthopyroxene such as hypersthene. Unfortunately, the two can look so alike that they are impossible to distinguish without a microscope. Norite typically occurs in small, separate intrusions, or as layers along with other mafic igneous rocks such as gabbro. Norite also formed in association with ancient basalt intrusions, beneath huge basalt dyke swarms. One famous norite intrusion is at Sudbury in Ontario. Here a cavity 30m/98ft deep has been excavated from solid norite to house the Neutrino Observatory to detect neutrinos, minute particles streaming from the stars. Norite is unusually low in natural radioactivity and acts as a shield to allow scientists to block out unwanted background radiation.

Identification: Norite is a dark grey rock with a slightly matted look dominated by quite long, prismatic black hypersthene or enstatite crystals. It looks very like gabbro, but the plagioclase feldspar tends to be sandy coloured, while in gabbro it is whiter.

Grain size: Phaneritic (coarse-grained), occasionally pegmatitic
Texture: Even-grained or porphyritic
Structure: Layering and xenoliths are common
Colour: Dark grey, bronze
Composition: Silica (58%), Alumina (17%), Calcium and sodium oxides (10.5%), Iron and magnesium oxides (11%), Potassium oxides (2%)
Minerals: Plagioclase feldspar (labradorite or bytownite); Pyroxene (hypersthene); Olivine; A little hornblende, biotite mica, quartz and alkali feldspar
Accessories: Magnetite, apatite, ilmenite, picotite
Phenocrysts: Plagioclase feldspar, hornblende
Formation: Intrusive: dykes, stocks, bosses, often with gabbro
Notable occurrences: Aberdeen, Banff, Scotland; Norway; Great Dyke, Zimbabwe; Bushveld complex, S Africa; Sudbury, Ontario

Anorthosite

Identification: Anorthosite is the lightest coloured of all the gabbroic rocks. Dark and light minerals are often aligned in long crystals, giving anorthosite a streaky look.

Anorthosite is almost entirely plagioclase feldspar. Over 90 per cent is either bytownite or labradorite. Labradorite crystals may show an iridescence known as labradorescence. Although not as abundant as basalt and granite, anorthosite often occurs in huge formations such as in Labrador in Canada, and in giant complexes such as South Africa's Bushveld along with gabbro and norite. It is also one of the rocks that makes up the Moon's surface. While lunar seas are basalt, highlands are anorthosite. When the Moon was young its surface was melted, not only from heat within but by meteor impacts. Light plagioclase feldspar floated to the top and, when the lunar surface cooled, it solidified to form anorthosite. The Earth has much less anorthosite, but it is found in ancient rocks. On the early Earth, anorthosite may have been as abundant as on the Moon, but Earth's surface is so dynamic that most has long since vanished.

Grain size: Phaneritic (coarse-grained)
Texture: Long crystals often aligned
Structure: Layering common
Colour: Light grey
Composition: Silica (51%), Alumina (26%), Calcium and sodium oxides (16%), Iron and magnesium oxides (5%)
Minerals: Plagioclase feldspar (labradorite or bytownite); Small amounts of pyroxene; olivine; magnetite and ilmenite
Formation: Intrusive: dykes (rare), stocks, batholiths, often with gabbro
Notable occurrences: Norway; Bushveld complex, South Africa; Sudbury, Ontario; Labrador; Stillwater, Montana; Adirondacks, New York; the Moon

Bushveld Complex
South Africa's Bushveld complex in the former Transvaal is one of the world's great geological wonders. It is by far the largest layered intrusion, covering up to 65,000km²/25,097 sq miles and reaching up to 8km/5 miles thick. It is incredibly rich in minerals, containing most of the world's chromium, platinum and vanadium resources, as well as a great deal of iron, titanium, copper and nickel. The whole complex formed in a remarkably short time about 2,060 million years ago, as magmas were poured out on the surface and intruded into the ground to form large, complex layers of gabbroic and mafic rocks such as norite, anorthosite and pyroxenite. Some geologists thought it was created by the hotspot above a mantle plume; others, noting the coincidence of dates with the nearby Vredefort meteor crater, thought it was a meteorite impact feature. A recent theory uses both ideas, suggesting a meteorite impact triggered off a mantle plume to fountain huge amounts of magma up from the mantle – almost as if the meteorite had burst the Earth's crust.

Essexite

Named after Essex County in Massachusetts where it occurs, essexite is the rock used in Scotland in its porphyritic form to make curling stones. It is the gabbro that forms when there is less silica in the melt. It is less viscous than gabbro and flows into small intrusions near the surface, cooling quickly to form medium- and fine-grained rocks. The lack of silica means that nepheline forms in essexite instead of quartz, so it is actually a foid rock, like foyaite and nephelinite. Essexite is also richer than gabbro in pyroxene, and its pyroxene is titanaugite.

Grain size: Medium to fine
Texture: Granular, sometimes porphyritic
Structure: Layering common
Colour: Light grey
Composition: Silica (45%), Alumina (15%), Calcium and sodium oxides (17%), Iron and magnesium oxides (18.5%), Potassium oxides (5%)
Minerals: Plagioclase feldspar (labradorite or anorthite); Pyroxene (augite); Biotite; Hornblende; plus small amounts of nepheline and alkali feldspar
Formation: Small intrusions, dykes, sills, often with gabbro
Phenocrysts: Augite
Notable occurrences: Lanarkshire, Ayrshire, Scotland; Kaiserstuhl, Baden, Germany; Oslo, Norway; Roztoky, Czech Republic; Tyrol, Italy; Essex Co, MA

Identification: Essexite is a fine- to medium-grained grey rock often with slightly larger dark spots of augite. It is attractively evenly mottled.

Troctolite

This is a gabbroic rock which has almost no pyroxene. Instead it is made of plagioclase feldspar and olivine, midway between anorthosite and peridotite. It very often occurs in layered igneous complexes such as South Africa's Bushveld, and, most famously among geologists, the Isle of Rhum in Scotland's Hebrides. Layered complexes are intrusions that seem a layer-cake of related igneous rocks formed within a single magma chamber in the Earth's crust. The Rhum complex formed some 60 million years ago, during the birth of the North Atlantic Ocean, when ancient north-west Europe and North America began to drift apart, allowing magma to flood on to the surface. The original magma was probably an olivine-rich basalt, but as minerals crystallized in the magma chamber, heavier minerals probably sank to the bottom, creating layers of troctolite on top of layers of peridotite. Troctolites elsewhere probably formed in a similar way.

Identification: In German, troctolite is called *Forellenstein*, which means 'trout rock', and the name is apt, for it looks just like the skin of a trout. The medium-grained dark grey plagioclase looks like the trout's scales. Olivine forms black spots within it which, just like a trout's spots, can be red, green or brown when wholly or partly altered to serpentine by exposure to the weather.

Grain size: Medium to coarse
Texture: Granular
Structure: Layering common
Colour: Grey studded with black, occasionally red or green
Composition: Silica (51%), Alumina (26%), Calcium & sodium oxides (16%), Iron & magnesium oxides (5%)
Minerals: Plagioclase feldspar (labradorite or anorthite); Olivine; Small amounts of pyroxene, magnetite and ilmenite
Formation: Intrusive: dykes, cone sheets, stocks, laccoliths, often with gabbro
Notable occurrences: Rhum, Scotland; Cornwall, England; Oslo, Norway; Harz, Germany; Wolimierz (Silesia), Poland; Niger; Great Dyke, Zimbabwe; Bushveld complex, South Africa; Stillwater, Montana; Oklahoma

DYKE, SILL AND VEIN ROCK

Fingers of magma ooze out into the country rock from every intrusion, either cutting across strata as dykes in which lamprophyres may form, or sliding between as sills to form rocks such as dolerites. As the intrusion begins to cool, cracks open in the solidifying rock. Residual fluids ooze into these cracks, altering the surrounding rock to greisens or solidifying to form veins of new rocks such as aplite.

Aplite

Aplites are unusually pale igneous rocks with fine, even grains that look just like unrefined sugar. They are closely related to pegmatites, and likewise form veins of crystalline igneous rock. Aplites, though, are fine-grained and tend to be much simpler in composition. In fact, they are basically quartz and potassium feldspar, with no mica, which is why they are so pale in colour. Moreover, while there are often complex zones of different composition in pegmatites, aplites are generally uniform throughout. Aplite veins form in almost every large granitic intrusion, striking like a pale scar across the host rock, and rarely more than a few centimetres across. As the intrusion cools and begins to crack, aplite veins develop when residual magma fills up the cracks. These aplites form at the lowest temperature of any igneous rock and water comes out of the melt as it loses pressure. So they crystallize very rapidly creating a texture that is remarkably fine-grained considering they form deep underground.

Identification: With its fine crystalline texture, aplite has a sugary look almost like pale sandstone. Unlike sandstone, though, the grains in aplite interlock, and there is none of the cement that glues sandstone grains together.

Grain size: Fine-grained
Texture: Even-grained or, occasionally, porphyritic
Structure: None
Colour: Pale pink or whitish
Composition: Silica (75% average), Alumina (14.5%), Calcium and sodium oxides (3.5%), Iron and magnesium oxides (5%)
Minerals: Quartz; Potassium feldspar (orthoclase or microperthite)
Accessories: Plagioclase feldspar, muscovite, apatite, tourmaline
Phenocrysts: Quartz, orthoclase feldspar, tourmaline
Formation: Aplite forms dykes in granite and granitic intrusions. It occasionally forms independent bosses, or on the rim of an intrusion
Notable occurrences: Wherever there are large granitic intrusions

Greisen

Identification: Greisen almost always occurs within granite, but is pale grey with almost no black mica.

Strictly speaking, greisen is a metamorphic, not igneous, rock, but it is always closely linked to granite, especially in tin-mining districts. In fact, it is granite that has been altered or 'metasomatized' by exposure to hydrothermal fluids or vapours rich in fluorine, lithium, boron and tungsten – just as basalt is metasomatized to spillite. As fluids flood through veins in granite, they alter the composition of the granite in the vein walls, destroying all the feldspar, and leaving just quartz and white mica. Gradually the vein becomes infilled with greisen. There is usually no definite boundary between the greisen and granite, and altered granite merges into unaltered granite imperceptibly. Greisens belong to the quartzolite family, the most quartz-rich of all rocks, with a quartz content of over 90 per cent.

Grain size: Medium-grained
Texture: Even-grained or, occasionally, foliated
Structure: None
Colour: Grey or brown
Composition: Silica (90% average)
Minerals: Quartz; White mica (muscovite, zinnwaldite, lepidolite, sericite)
Accessories: Topaz, fluorite, apatite, tourmaline, rutile, cassiterite, wolframite
Formation: Short vein infillings no more than a few hundred metres long
Notable occurrences: Skiddaw (Lake District), Cornwall, England; Galicia, Spain; Fichtelberg, Erzgebirge, Germany; Portugal; Queensland, New South Wales, Tasmania, Australia

Lamprophyres

Lamprophyres get their name from the Greek for 'glistening mixture', and it is apt for these rocks are stuffed with large, gleaming phenocrysts of mica, amphiboles and olivine, giving the rock a very distinctive appearance. Unusually, they have no feldspar phenocrysts whatsoever; all their feldspar is in the fine groundmass. They are dark, or even ultramafic, rocks, and probably form from cool melts of metasomatized (altered) mantle material. They are the classic dyke rocks. Most dyke rocks are simply versions of the same rocks that form larger intrusions. Lamprophyres alone form almost exclusively in dykes – though in recent years small lava flows and plutons have been found. They typically occur in dykes near tonalite and granodiorite plutons. The lamprophyres are a very varied group, and some geologists think they should be called a facies – a diverse group of rocks that simply crystallized under similar conditions. The most widespread form is minette.

Camptonite: Named after Campton in New Hampshire, this dark lamprophyre has a hornblende, labradorite feldspar and pyroxene groundmass with phenocrysts of the amphiboles kaersutite and ferrohornblende, along with titanaugite, olivine and biotite.

Vosgesite: Named after the Vosges in Alsace, France, vosgesite has phenocrysts of hornblende along with augite and olivine. It is one of the calc-alkaline lamprophyres normally found along with rhyolites and basalts in island arcs and subduction zones.

Grain size: Mixed
Texture: Porphyritic
Colour: Dark grey with dark, even black phenocrysts
Composition: Variable. Minette: Silica (47.5% average), Alumina (9.3%), Calcium and sodium oxides (11.5%), Iron and magnesium oxides (26%).
Groundmass: Plagioclase feldspar, feldspathoid, carbonates, monticellite, mellilite, mica, amphibole, pyroxene, olivine, perovskite
Phenocrysts: Biotite/phlogopite, amphibole (hornblende, barkevikite, kaersutite), pyroxene (augite), olivine
Formation: Mainly dykes
Notable occurrences: Cairngorms, Cheviots, Scotland; Lake District; England; Ireland; Vosges, France; Black Forest, Harz, Germany; Wasatch, Utah

Dolerite

A tough stone used for road metal, dolerite is a dark, mafic rock, the medium-grained equivalent of basalt and gabbro. It is known in the United Kingdom as diabase and is typically found in sills, such as New Jersey's vast Palisades sill. The famous bluestones of England's ancient stone circle Stonehenge were carved from dolerite cut from sills in the Prescelly Mountains of Wales. Usually when magmas cool and crystallize, dark minerals such as olivine crystallize first, followed by feldspar and mica, leaving quartz and any other silica to fill in the gaps. Lath-like crystals of feldspar form first, and the dark minerals are forced to fit in between them – often growing right around them in what is called ophitic texture.

Palisades Sill
New Jersey's Palisades are a dramatic line of brown cliffs that tower anything from 107m/350ft to 168m/550ft above the west bank of the Hudson River. The cliffs are the exposed margin of a vast sill of diabase (dolerite) rock 305m/1,000ft thick and 72km/45 miles long that dips away westwards. Radiation dating has shown the sill formed between 186 and 192 million years ago in the Early Jurassic, when a fat wedge of magma squeezed between layers of sandstone and shale. As the magma cooled and solidified, it cracked into the columns that characterize the cliff face – and earned the Palisades their name, given by explorers with Verrazano in 1524, who thought the cliffs resembled the forts of wooden stakes built by local Indians. In the 19th century, these rocks were ruthlessly exploited for building stone, and many a New York sidewalk is made of 'Belgian stone' from the Palisades. Eventually, in the 1930s, the area was designated an Interstate Park to halt the destruction.

Grain size: Medium
Texture: Often ophitic, with large clinopyroxene (augite) crystals enclosing plagioclases
Structure: Vesicles and amygdales common
Colour: Dark grey, black, with green tinge when fresh. May be mottled white.
Composition: Silica (50%), Alumina (16%), Calcium and sodium oxides (13%), Iron and magnesium oxides (18%)
Minerals: Plagioclase feldspar (labradorite); Olivine; Pyroxene; Biotite; Magnetite; Ilmenite; Quartz; Hornblende
Phenocrysts: Olivine and/or pyroxene or plagioclase
Groundmass: Plagioclase and pyroxene with olivine or quartz
Formation: Sills and dykes, often in large dyke swarms; occasionally lava flows
Notable occurrences: Whin Sill, NE England; Lake Superior, Canada; Palisades, New Jersey

Identification: Dolerite is best identified by its dark, greenish colour and its medium grain size.

PEGMATITES

Pegmatites are the cream of igneous rock, the places where all the biggest crystals and rarest minerals are concentrated when an igneous intrusion reaches its final stages of crystallization. Pegmatite formations are typically small pods and lenses no bigger than a house, but they are the source of some of the world's best gems and most valuable minerals.

Pegmatite features

Identification: Pegmatites are instantly recognizable from their gigantic crystals. What is harder to identify is the particular crystals within them, and so the particular kind. Large creamy or pink crystals are usually feldspar, white sugary crystals may be quartz, brown striped crystals are mica, black may be tourmaline. More colourful crystals are rarer minerals.

Quartz

Pink beryl

Amazonite feldspar

Lithium pegmatite: Many pegmatites are enriched with the mineral lithium, turning mica to lepidolite and creating the spodumene gems – lilac kunzite and green kunzite.

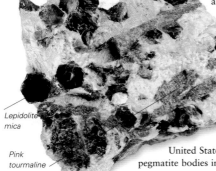

Lepidolite mica

Pink tourmaline

Pegmatites are perhaps the most fascinating bodies of rock in the world. No other rock formation contains such a wealth of large, spectacular crystals. All the world's largest natural crystals have been found in pegmatites. Even the average grains in pegmatites are not just clearly visible, as in coarse-grained granite, but substantial – at least the size of a grapefruit. Some pegmatite crystals are truly gigantic. Tourmaline and beryl crystals the size of a log are often found, while a spodumene crystal found in a pegmatite in South Dakota in the USA was a gigantic 13m/42ft long.

Pegmatites are also the source of an amazing range and variety of minerals. Over 550 different kinds of mineral have been found in pegmatites. Pegmatites are the source of many of the world's gems. Besides fabulous topaz and wonderful garnets, gems of all the beryl varieties (aquamarine, morganite, golden), all the tourmalines (pink, green, and multi-hued elbaite) and all the spodumenes (kunzite and hiddenite) are found in pegmatites. Pegmatites are also sources of rare elements such as beryllium, niobium, tantalum, rubidium, caesium and gallium, as well as tin and tungsten. Because they are so richly concentrated here, pegmatites are even major sources for more common minerals such as feldspar and quartz.

The term pegmatite was originally used in the early 19th century to describe graphic granites, which often occur in pegmatites. Now the term is used to describe any small body of igneous rock with crystals at least 1.3cm/0.5in across. They vary hugely in size and shape. Some are veins. Some are shaped like lenses. Some are shaped like knobbly turnips. The smallest are typically no bigger than a mattress but a few giant pegmatites are 3.2km/2 miles long and 0.5km/0.3 mile wide. Pegmatites are by no means isolated structures. In the famous Black Hills district of South Dakota in the United States, there are an estimated 24,000 pegmatite bodies in an area of 700km²/270 sq miles!

Grain size: Very coarse-grained. Crystals are at least 1–2cm/0.5in across, average 8–10cm/3–4in, and can be much bigger.
Texture: Hugely varied, with complex zoning and lots of vugs (open cavities)
Structure: None
Colour: Pale pink or whitish
Composition: Silica (75% average), Alumina (14.5%), Calcium and sodium oxides (3.5%), Iron and magnesium oxides (5%)
Minerals: Quartz; Potassium feldspar (albite and perthite)
Accessories: Plagioclase feldspar, muscovite, apatite, tourmaline, plus minerals containing elements such as tin, tungsten, niobium, tantalum, beryllium, gallium, rubidium and caesium
Large crystals: Quartz, feldspar, tourmaline, beryl (aquamarine, morganite, golden), spodumene (kunzite and hiddenite), tourmaline
Formation: Dykes, veins, pods, lens around the margins of intrusions, Pegmatites are typically either granite or syenite
Notable occurrences: Pegmatites occur all around the world, wherever there are granite and syenite intrusions. Famous examples include those on the Island of Elba in Italy, Madagascar (especially Anjanbonoina), Pakistan and the Mesa Grande, California (notably the Himalaya Mine in Pala County). They are most abundant in mountain chains and stable continental shield areas, such as the Canadian Shield, Greenland and north Russia. Shield pegmatites are usually at least a billion years old. Mountain chain pegmatites such as those in the Himalayas are no more than 5 to 20 million years old.

Pegmatite formation

Pegmatites usually form at the margins of a large pluton, clustering like currents on its surface, or extending like fingers into the mass, or outwards into the surrounding country rock. Occasionally they are found completely separated from their parent intrusion in pockets in the country rock. It is thought that pegmatites form in the last stages of the crystallization of an intrusion. With the main body of rock formed, the bulk of the common minerals have already crystallized, leaving just a few small pockets to be filled in to create pegmatites. The melt left to create them is rich not just in rarer elements such as boron and fluorine, but also volatile liquids – and a great deal of water. It is this high water content that allows the crystals in pegmatites to grow so big. All this water dramatically increases the mobility of particles within the melt, and this means they can travel farther and faster as they become incorporated into crystals. Normally, large crystals form only when magmas cool very, very slowly from high temperatures, but the water in pegmatites means huge crystals grow rapidly at temperatures of no more than 100–200°C/212–392°F. Pegmatites can form from just about any kind of igneous intrusion including gabbro and diorite – and even in metamorphic gneisses and schists – but most form from granites and syenites and so have the same basic ingredients – quartz and potassium feldspar with a little muscovite mica. Pegmatites can be divided into simple and complex. Simple pegmatites are basically very coarse-grained equivalents of the parent rock, made of the same three basic ingredients, plus a little tourmaline. Complex pegmatites form later and have higher concentrations of rare minerals. Lithium, for instance, is typically found at concentrations of 30ppm (parts per million) but in complex pegmatites, lithium concentrations can reach over 700ppm. Complex pegmatites are typically divided into highly complex zones, with graphic granite in one place, tourmalines in another, and so on. They are often riddled with open cavities. It is the complex pegmatites that are the source of the most valuable and spectacular crystals.

Black tourmaline

Tourmaline pegmatite: Tourmaline is a boron mineral and tourmaline pegmatites are created when the boron content is especially enriched in the last stages of the mix. An increase in fluorine often creates the gem topaz.

Varieties of pegmatites: Pegmatites are very varied. Some are named after the main source rock. Others are named after a mineral or element that is particularly enriched in them.

Rock source pegmatites:
Granitic pegmatite
Syenitic pegmatite
Gabbroic pegmatite
Diorite pegmatite

Mineral- or element-enriched pegmatites:
Tourmaline pegmatite
Lithium pegmatite
Beryl pegmatite
Emerald pegmatite
Spodumene pegmatite
Albite pegmatite
Quartz-albite pegmatite
LCT pegmatite

Rock source and element-enriched pegmatites:
Phosphate granitic pegmatite
Boron granitic pegmatite
LCT granitic pegmatite

LCT pegmatites: LCT (Lithium-Caesium-Tantalum) pegmatites are enriched not just with the elements lithium, caesium and tantalum, but also with rubidium, beryllium, gallium and tin. Granitic LCT pegmatites are host to many of the world's most precious gemstones, including emerald, chrysoberyl and topaz from Minas Gerais in Brazil, sapphire and ruby from Afghanistan and Pakistan, and gem tourmaline.

Isola d'Elba, Italy

The western end of the Island of Elba off the west coast of Italy is riddled with pegmatites that have been one of the world's richest sources of beryls and tourmalines, including the elbaite variety named after the island. In 1805, the first quarries were dug to extract the local granite to build houses and roads. Then in 1820, a local mineralogist called Captain Foresi noticed large, colourful crystals in the rock, later found to be tourmalines. In 1830, Foresi opened the first tourmaline mine at Grotta d'Oggi, or Cave of Today. He then traced the pegmatite zone and discovered many other sources of tourmalines and beryls such as the Masso Foresi (Foresi's Mass), the Fonte del Prete and la Speranza. Scores of quarries and mines were opened up and yielded many thousands of fine tourmalines and beryls. However, these localities existed within around 10 sq km (six sq miles) of each other and, unsurprisingly, by the end of the 19th century, the existing veins – sunk deep into granite rock – had been mined to exhaustion, and the best stones taken away. Many are now on display in mainland Italy, at the Florence Mineral Museum. Visitors to the island do occasionally still find fine elbaite crystals here, and many collectors believe there are countless new pegmatite-rich veins waiting to be discovered. However, with new laws to safeguard the landscape of the island against future mining, it is unlikely that the Isola d'Elba will regain its previous status as a mecca for those seeking pegmatite gems.

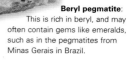

Smoky quartz

Feldspar

Yellow beryl

Beryl pegmatite: This is rich in beryl, and may often contain gems like emeralds, such as in the pegmatites from Minas Gerais in Brazil.

SEDIMENTARY ROCKS: LUTITES

Sedimentary rocks are made from sediments of loose material that is gradually lithified (consolidated and turned to stone) over millions of years. Many sediments are fragments, or 'clasts', of older rocks broken down by weather. Clasts are mainly silicates (quartz, feldspar, mica). Rocks made from them are described as siliciclastic, and classified according to grain size. The finest grained are lutites or mudrocks, made mainly of clay- and silt-size grains. They include claystone, mudstone, siltstone and shale.

Claystone and mudstone

These siliciclastic rocks are the most abundant sediments on Earth. Half of all sedimentary rocks are clays and muds, and beds of clay stripe nearly every sedimentary formation. London and Paris both sit on vast dishes of clay that provide the bricks that built the cities and the rich soils of the farmlands around. Dull, flat and common claystones may be, but they are the most useful of all rocks. Impure clays are used for making bricks and tiles – and their organic content makes many self-firing. Pure clays such as kaolinite are so wonderfully mouldable they still provide the best materials for making pottery as well as fillers for papers.

No rocks are made from tinier grains than these. More than half the grains in claystone are clay-sized – less than 4µm/0.15mil across. Over two-thirds of the grains in mudstones are clay-sized. Tiny grains like these are the last remnants of rocks broken down by weathering. Being so light they are carried farthest from their source. When rivers flow into the sea and drop their sediment load, clays fall last. Many grains float right out into the deep ocean before sinking to become part of the ocean floor ooze.

Most claystones form from clay sediments that settle in shallow waters just off shore, in calm areas below the waves. Clays like these are dotted with fossils of sea creatures, from tiny shellfish to giant marine dinosaurs. Besides these marine sediments, claystones may also form on lake beds and where rivers flood. Some claystones are not sediments at all but 'residuals' developed as rocks are altered in situ to create soils such as laterites. Few claystones are old, however. Buried under later sediments, they are quickly consolidated first into shale and then to slate in a process called diagenesis. So clay tends to appear only in younger geological formations.

Clay minerals are divided into four broad groups: the kandites (such as kaolinite) formed by the breakdown of potassium feldspar; the illites formed from feldspars and mica; the smectites (including montmorillonite) formed from pyroxenes and amphiboles; and the chlorites. Each claystone contains its own mix of these four, with varying proportions of organic material as well. Illites and montmorillonites are most prevalent.

Black mudstone: Mudstones are defined as rocks made of one-third silt grains and two-thirds ultra-fine clay grains. They are basically hardened mud, and like most muds they are often rich with both plant and animal matter. It is this organic matter that often turns them black.

Claystone: The particles in clay are so tiny that, when pure, it feels smooth and slippery when wet, like plasticine. Claystones look like earthenware and come in many colours from grey clays rich in plant material to red clays rich in iron oxide.

Grain size: Over 50% of grains are clay-sized, less than 4µm/0.15mil
Texture: Even-grained, with fossils. No grittiness like silt. Can be plastic and often sticky when wet.
Structure: No fine layering like shale, but clay beds show larger scale stratification, including originally horizontal topset and bottomset beds formed on the top and beyond a delta, and originally sloping offset beds formed on the delta front. Sun cracks, rain prints etc are common. All clay particles are microscopically layered, which makes clays plastic and slippery when wet as layers slide over each other. Mudstones have a blocky, massive fabric.
Colour: Black, grey, white, brown, red, dark green or blue
Composition: Mix of detrital quartz, feldspar and mica. Iron oxides turn clays red or brown. Organic matter turns them black.
Different groups of minerals: Kandites such as kaolinite; Illites; Smectites such as montmorillonite; Chlorites
Formation: From clays and muds settling offshore, on lake beds and on river floodplains. Also as residuals as rock is altered in situ.
Notable occurrences: London Clay (London basin), Oxford Clay (Weymouth to Yorkshire), England; Paris basin, France; North German basin, Molasse basin, Upper Rhine, Germany; Kamchatka, Russia; Sydney basin (New South Wales), Australia; Trinity River, Texas; Appalachian Mountains; Newland, Montana; Muldraugh Hill, Kentucky

Shale

Gray shale: Shale is often dark grey or brown with a thin, platy structure and no visible grains.

Like claystone and mudstone, shale is made from the fine particles that settled on the floor of shallow seas and lakes long ago. Yet unlike claystone and mudstone, shale has laminations like the pages of an ancient book, created as it was squeezed by the weight of overlying sediments. As a result, shale looks flaky like slate, and likewise splits easily into thin layers, a tendency called fissility. The layers vary from paper thin to card thick. Unlike slate, though, shale often contains the fossilized remains of sea life, buried in the mud and preserved forever, albeit somewhat flattened, as the mud turned to stone. Shale varies tremendously in colour according to the minerals it contains. Black shales are rich in carbon from organic remains (typically plankton and bacteria), which often turned to kerogen as the rock formed. Oil shales are black shales so rich in kerogen and bitumen (at least 20 per cent) that they yield oil if heated intensely. On average, 1 tonne/0.98 tons of rock can yield 750 litres/ 165 gallons of oil. Scientists have yet to find a way of extracting this oil economically.

Black shale: This shale is black because the remains of sea creatures it contains were never oxidized. Some formed in basins in which circulation was restricted. Others may have developed in times when global warming peaked, cutting down circulation of deep ocean currents. Black shales are famous for the extraordinary preservation of fossilized sea creatures. Even soft tissues often leave impressions.

Grain size: 50% of grains are clay-sized, less than 4μm/0.15mil
Texture: Even-grained, with fossils. No grittiness like silt.
Structure: Splits easily into thin layers
Colour: Black, grey, white, brown, red, green or blue
Composition: Mix of detrital quartz, feldspar and mica. Iron oxides turn shales red or brown. Organic matter turns them black.
Formation: From clay sediments settling offshore, on lake beds and on river floodplains then undergoing diagenesis
Notable occurrences: Many locations around the world: *Black shale*: Posidonia, Hünsruck, Germany; Chattanooga, TN; New Albany; PA; Great Plains of Kansas, Oklahoma, South Dakota; *Oil shale*: Torbane Hill, Scotland; Estonia; Lithuania; Israel; Tasmania; Green River, Colorado

Jurassic mud

It was in the remarkable Oxford and London Clays of England that Victorian fossil hunters made many of the great finds that led to the discovery of dinosaurs and many other prehistoric creatures. Both contain a wealth of marine fossils, but the Oxford Clays are especially rich. The sediments they are made from developed some 140–195 million years ago in the Jurassic period. At this time, southern England was entirely covered by the waters of a tropical ocean, teeming with sea creatures. Oxford Clay is full of the fossils of fish and shellfish that swam there then, including the giant Leedsichthys fish, and countless ammonites and belemnites. But it is most famous for its marine dinosaurs, including the plesiosaurs *Cryptoclidus*, saucer-eyed *Opthalmosaurus* (pictured above), and the awesome *Liopleuridon*. At almost 25m/80ft long, *Liopleuridon* was the biggest carnivore that ever lived, with a mouth 3m/10ft long and teeth twice as long as those of the *Tyrannosaurus rex*.

Siltstone

Siltstones are much less common than clays and muds and rarely form thick beds. At least half the grains in siltstone specimens are coarser, silt-sized grains, 4–60μm/ 0.15–2.5mil across – large enough to be visible with a magnifying glass. They are mostly quartz, making the rock tougher than clay, and giving it a slightly gritty feel. Because the grains are heavier, siltstones form closer to the shore than clays, and often show ripple marks and crossbedding created by the interplay of the river currents and waves in shallow water. As the flow of water changed with the seasons, so sediment deposition varied. So pale siltstones are often found interlayered with darker mudstones.

Grain size: Over 50% of grains are silt-sized, in the range 4–60μm/0.15–2.5mil
Texture: Even-grained, with fossils. Slightly gritty.
Structure: Often laminated. Shows crossbedding and ripple marks.
Colour: Pale grey to beige
Composition: Mix of detrital quartz, feldspar and mica
Formation: From clay sediments settling in river deltas, on lake beds and on river floodplains
Notable occurrences: Many locations around the world including south-east England, The Great Plains of North America, China

Identification: Siltstone is easily identified by its pale colour, just-visible grains and slightly gritty feel. It is often laminated with bands of dark mudstone.

MORE MUDROCKS

Besides claystones, mudstones and siltstones, there is a wide range of other mudrocks, including marls,
bentonites and boulder clays. Marls are earthy, crumbly mixtures of lime and silicate fragments.
Bentonites are clays formed from volcanic ash that falls on the sea bed. Boulder clays are basically
the debris left behind by moving sheets of ice.

Marl and marlstone

Green marl: Marls are often given a green tinge by the potassium mica mineral glauconite. These green marls are often very rich in fossils. There are extensive deposits in places such as England's Isle of Wight and North America's Atlantic seaboard.

Red marl: The lime and clay content of marl means that they are normally white, grey or brown, but some marls are very rich in iron, which turns them red. Strictly speaking, they are not marls, since they have a low lime content, but they have the same earthy texture.

Since many sediments form on sea and river beds, it is inevitable that they become enriched with a fair share of debris of shellfish and other marine life. As the rocks form, this debris turns to carbonates, first to calcite and aragonite, and eventually to calcite and dolomite in older rocks. Limestones and chalks are almost pure calcium carbonate or lime. Mudstones and claystones, however, contain only a little lime. Marlstone lies in between, rich in both lime and silicate fragments of weathered rocks.

Strictly speaking, marlstone is the rock, while marl is the soft, earthy material that forms as this and other rocks are weathered, but geologists often use the word marl as a general term for any hybrid of mudstone and fine-grained limestone. Extra lime turns marls into limestones; less turns them into clays and mudstones.

The mixed lime and clay content makes all marlstones soft and friable, even when they are not actually earths. Many disintegrate in water, and the lime content means they are easily dissolved in dilute hydrochloric acid or even vinegar.

In some marls, called shelly marls, the carbonate material is actual shell fragments. Shelly marls like these are much valued by farmers as a source of lime, because the lime is easy to extract. In others, the lime is a fine powder mixed in completely with the quartz and feldspar grains. Marls that form in freshwater are quite similar to those that form in the sea. They also often contain shell fragments, but most of the organic material usually comes from algae.

In England, there is a group of rocks called New Red marls that form beds up to 300m/1,000ft thick in places, as part of the Keuper system. These are iron-rich clays rather than pure marls, because they contain only a little calcium carbonate. They probably formed in salt lakes in desert conditions, and in places contain thick salt beds, such as those in Cheshire. Some hard slates in Germany are also described as marls, including the important copper-bearing marl-slates of the Mansfeld area.

Grain size: Over 50% of grains are clay-sized, less than 4µm/0.15mil

Texture: Earthy, even-grained, with fossils. Marl has none of the grittiness of silt, feeling much softer. The lime is sometimes powdery, sometimes in the form of shell fragments.

Structure: No fine layering like shale, but clay beds show larger stratification. Sun cracks, rain prints etc are common. All clay particles are microscopically layered, which makes clays plastic and slippery when wet as layers slide over each other. The high lime content makes these rocks very friable and crumbly. The combination of lime and clay makes a very good basis for soil, which is why marl is often added to soils to improve fertility.

Colour: Various, including brown, white or grey; may also be red with iron content or green with glauconite

Composition: Even mix of carbonates (mainly calcite) from organic sources and detrital quartz, feldspar and mica

Formation: From clays and muds settling offshore, on lake beds and on river floodplains

Notable occurrences: North Yorkshire, Leicester, Northamptonshire, Oxford, Exmouth, Vale of Eden, England; Valkenburg, Netherlands; Paris basin, France; Mansfeld, Bavaria, Germany; Sydney basin (New South Wales), Australia; Weka Pass, Canterbury Plains, New Zealand; Green River Formation, Wyoming; South Dakota; Atlantic coastal plain (New Jersey, Delaware, Maryland, Virginia)

Bentonite

Named after a kind of clay found near Fort Benton, Wyoming, in 1890, bentonite is a clay formed by the alteration of volcanic ash that has settled on the sea floor. Similar clays called tonsteins form when ash is weathered in the acidic waters of coal swamps. Bentonites and tonsteins consist of mostly smectite clays, but also contain unchanged volcanic fragments such as quartz grains and mica flakes. They may also contain beads of volcanic glass. There are two kinds of bentonite: sodium bentonite and calcium bentonite. Sodium bentonite is an incredibly useful material because it swells enormously when wet, creating a gelatinous mass that has been used for everything from sealing dams and drilling for oil to cat litter and detergents. Calcium bentonite makes the absorbent clay called fuller's earth. Bentonites are typically found interbedded with shallow marine limestones and shales, and represent a sudden and dramatic event when a volcano showered the sea floor with huge quantities of ash. Although beds up to 15m/45ft thick have been found, most are less than 0.3m/1ft. Although they are frequently altered beyond recognition, they can give a valuable insight into past volcanic events. In the Ordovician period, much of the eastern USA was covered by an immense ashfall 1m/3ft thick, leaving extensive bentonite deposits from Tennessee to Minnesota.

Bentonite: Bentonite looks like clay mud, but is generally and buff to olivegreen colour. If it absorbs water it will swell dramatically.

Grain size: Over 50% of grains are clay-sized, less than 4µm/0.15mil
Texture: Earthy, even-grained. Slippery and plastic when wet. Greasy or waxy feel.
Colour: White to light olive green, cream, yellow, earthy red, brown and sky blue. Bentonite turns yellow on exposure to air.
Composition: Smectite clay minerals, quartz grains, mica flakes, volcanic glass beads, calcite and gypsum
Formation: Bentonite formed by the alteration of volcanic ash falling on the sea bed. Tonstein formed from ash falling in coal swamps.
Notable occurrences: Redhill (Surrey), Woburn (Bedfordshire), Bath (Avon), England; Spain; Italy; Poland; Germany; Hungary; Romania; Greece; Cyprus; Turkey; India; Japan; Argentina; Brazil; Mexico; Saskatchewan; Wyoming; Montana; California; Arizona; Colorado; Black Hills, South Dakota

Marl, the farmer's friend
Farmers have added marl to soil to improve its fertility for thousands of years. When added to acid soils, the lime in marl helps to neutralize the acidity. It also helps to glue sand grains together so that they retain heat and water better. When added to clay soils, wonderfully it has the opposite effect – helping to make the soil more crumbly and friable and allowing air, heat, water and roots to penetrate better. So marl promotes plant growth in a number of ways: it increases the food available for plants, and makes it easier for them to reach it. For centuries until artificial fertilizers began to take over, marl was dug from marlpits in large quantities. In the eastern USA, where marl is abundant, farmers often put 20–30 tonnes/19.7–29.5 tons of it on every 0.4 hectare/1 acre of land in the 19th century, giving magnificent potato, tomato, and berry crops. Clover grew especially lush on marled soils.

Boulder clay

Also known as tills and ground moraine, boulder clays are a legacy of the great ice ages that once covered much of northern Europe and North America in vast ice sheets. Mixed in like a fruit cake are large pebbles and boulders that were swept along beneath the glaciers and fine clay from rocks shattered by frost and stripped by moving ice. The materials in the boulder clay reflect where the ice travelled. Thus in Britain, boulder clays near Triassic and Old Red Sandstone areas are red, while near Silurian rocks they are buff or grey, and those near chalk can be white.

Although the biggest boulder clay deposits were left by past ice ages, they are forming even today under glaciers and ice sheets in polar and mountain regions.

Grain size: Mixture of pebbles, boulders and clay-sized grains, less than 4µm/0.15mil. Boulders can weigh up to several tons.
Texture: Smooth clay embedded with angular stones
Colour: Varies according to original rock – red, white, grey, brown, black
Composition: Depends on the original rock
Formation: Debris accumulated and swept along beneath glaciers and ice sheets
Notable occurrences: All across northern Europe and northern North America, especially East Anglia in England and the North German plain

Identification: With large stones and boulders set in a sticky clayey mass, boulder clay is unmistakable. The interesting task is to work out the origin of the material.

SANDSTONES

Sandstones are second only to mudrocks in abundance, making up 10–15 per cent of all Earth's sediments, and because they are so durable, they often form some of the most prominent hills and landmarks, as well as providing valuable building stone. They are made mostly of sand-sized grains 60μm–2mm/ 2.5–80 mil across. At least half the grains must be this size for it to be classed as a sandstone.

Sandstone

Sandstone, as its name suggests, is made from grains of sand – quartz, feldspar, or simply sand-sized fragments of rock. Sometimes the sand was piled up by desert winds, and the grains were worn almost round as they were buffeted along. Sometimes the sand was laid down on river beds, beaches or in shallow seas, and the grains are a little more angular. The sharpest sand of all came from glacial debris or high up in rivers, where it had not travelled far.

Beach sand is typically yellow, but each sandstone is stained by the cement that binds the sand. Limonite cement gives some sandstones a yellowish hue. Calcite turns them white – perfect for glass. Bitumen turns them black like the sandstones of Alberta, while iron oxides stain them red and brown. It is these warm reds and browns that are seen in New York's famous brownstone fronts, and the rusty red mesas and buttes of Utah, Colorado and Arizona. Occasionally, the cement is so weak that the rock crumbles in your hands. Most of the time, it is much harder, and sandstones resist erosion to create some of the world's most dramatic landscapes – high ridges, steep bluffs and towering tablelands. Sandstone's toughness also makes it the perfect building stone, more widely used than any other.

Sandstones can be like a 'book' displaying the history of their formation. Cracks reveal where sand dried out in the sun. Ripples show where waves rolled over it. Bedding marks bear witness to the way sand deposition changed continually season by season, year by year. Desert sandstones, like those in Zion Canyon, Utah, may even capture the shape of ancient wind-blown sand-dunes. Most sandstones are also rich in fossils of the creatures that burrowed in the sand, or lived in the waters above it. In the brownstone quarries of Portland, Connecticut, huge footprints have been found that were made by dinosaurs that walked over these sands long ago.

Identification: With their visible grains of sand, sandstones are easy to identify but it is much harder to distinguish the kind of sandstone. In a fresh surface, you may be able to identify quartz and feldspar. Quartz grains are milky to clear, glassy with no cleavage marks. Feldspars are usually white or pinkish, with marked cleavage planes. They may be dissolved out leaving holes or changed to clay.

Malachite sandstone: Sandstones are never pure quartz sand or even quartz and feldspar. Most contain traces of minerals. This sandstone from the Triassic period is specked with green malachite.

Grain size: Over 95% sand-sized grains, 60μm–2mm/ 2.5–80mil
Texture: Gritty texture like solid sand. Grains well sorted, often well rounded. The amount of cement between grains varies widely.
Structure: Typically occurs in blanket-shaped deposits varying from a few metres to several hundred metres thick. Sandstones are usually interbedded with mudstones, limestones and dolomites. They usually display dramatic cross-bedding and ripplemarks, reflecting their formation in high-energy environments. Aeolian rocks often show sand-dune shapes.
Colour: Variable – typically red, brown, greenish, buff, yellow, grey, white
Composition: 40–95% quartz, with feldspar and rock fragments. Other components include mica, clay, organic fragments, plus many heavy minerals. Quartz and calcite cement.
Formation: Nearly all sedimentary environments from small alluvial fans to vast deep-sea plains. Some form in high-energy marine environments such as beaches. Some are formed from aeolian (wind-blown) sand-seas in deserts, where there is a ready supply of sand. A type called ganisters forms when most other kinds of grain are leached away, leaving just quartz sand.
Notable occurrences: Western Highlands, Scotland; Pennines, England; Apennines, Italy; Carpathians, Romania; Nile Valley, Egypt; India; Appalachian Mountains; Colorado and Allegheny plateaux; Montana

Old Red Sandstone

The Old Red Sandstones are among the most famous and studied of all rock formations. They are a gigantic sequence of rocks that formed from sediments piled up in a vast basin that stretched across what is now north-west Europe in the Devonian period from 408 to 360 million years ago. This was the time in the Earth's history when the first fish swam, the first land plants grew and the first insects crawled. As the massive Caledonian mountain range was slowly worn away, its remnants accumulated in this basin and in time turned to stone. So extensive were these sediments that geologists used to refer to them as the Old Red Sandstone Continent – and included the Catskill Mountains of North America in it, although these actually formed quite separately at roughly the same time. The Devonian sediments are by no means all sandstones, nor did they all form in the same way. Some were laid down by rivers, some in the sea and some in lakes. But the dominant beds are the massive layers of red sandstone. The basin lay in the baking tropics south of the Equator in Devonian times, and these sands piled up in desert sand-seas and alluvial fans. They acquired their distinctive red colour as moisture later rusted iron in them.

Identification: Old Red Sandstone formations are best identified by their location and the fossils they contain, such as the famous Devonian fish. These formations include shales and other mudrocks as well as sandstone. The sandstone can be recognized by its visible sand grains, often stained red by iron oxides.

Grain size: Over 95% sand-sized grains, 60µm–2mm/2.5–80mil
Texture: Gritty texture like solid sand. Grains well sorted, often well rounded. The amount of cement between grains varies widely.
Structure: See sandstone
Colour: Red, green, grey
Composition: See sandstone
Formation: Most of the red sandstones formed in vast alluvial fans spilling into desert basins, and in desert sand-seas
Notable occurrences: Shetland, Caithness, Midland Valley, Borders, Scotland; Fermanagh, Antrim, Northern Ireland; Mid-Wales; Shropshire, Devon, Somerset, England. North American red sandstones: western Canada; Catskill Mountains

Brownstone fronts
In the decades after the American Civil War in the 1860s, it was the fad to clad well-to-do houses in Boston and especially New York with brownstone. Entire districts became characterized by their brownstone fronts. Brownstone is a feldspar-rich sandstone that formed about 200 million years ago in the Triassic period. Iron-oxide cement coloured it a warm chocolate-brown. The most distinctive formation is near Portland in Connecticut, and in the late 1800s, the Portland quarries boomed. Yet the fad for brownstone did not last long – partly because of the coming of concrete, and partly because brownstone, which formed in horizontal layers, was set in buildings vertically face out or 'face-bedded.' Face-bedded like this, the brownstones quickly flaked off as water got in behind the layers and froze. Now these beautiful old buildings are cherished again, and the fronts are being restored with fresh stone from the reopened Portland Quarry and better mortar.

Greensand

This can be either a sandstone or mudrock turned green by tiny pellets of the clay mineral glauconite, which gets its name from the Greek for 'blue-green'. Some greensands are over 90 per cent glauconite with just a little quartz sand and clay. Glauconite is a potassium iron aluminium silicate and is a useful potash fertilizer and valuable water softener. It forms in shallow seas (50–200m/164–656ft deep) only at times when sediments are piling up slowly, allowing sealife to burrow widely. Pellets form as faeces or the insides of dead foraminifera shells are chemically altered. However, glauconite-rich rock is not always green because it is turned brown or yellow by weathering. Most greensands were laid down in the Jurassic and Cretaceous periods. In England, sandstone beds formed at this time are called greensand whether they contain glauconite or not.

Grain size: Clay to sand size
Texture: Sometimes gritty and sandy, sometimes smooth like clay
Structure: See sandstone
Colour: Red, green, grey
Composition: Mostly glauconite, with quartz sand and clay
Formation: Greensands form on shallow, slowly sedimenting sea beds. They are often the last stage in a sedimentation sequence and typically appear just below an unconformity.
Notable occurrences: The Weald, Dorset, Berkshire, Oxfordshire, Bedfordshire, England; Boulonnais, France; New Jersey; Delaware

Identification: Unweathered, greensand is coloured pale olive green by glauconite, but this turns brown or yellow when exposed to air, and so is less distinctive.

ARENITES AND WACKES

There are two main kinds of sandstone: arenites and wackes. Arenites are all sand-sized grains (60μm–
2mm/2.5–80mil) with little cement. Wackes are less sorted, with sand chaotically embedded in silt and
clay. Both arenites and wackes may be mostly quartz (like orthoquartzite), a mix of quartz and feldspar
(like the arenite arkose), or 'lithic' – that is, made of various rock fragments (like the wacke greywacke).

Orthoquartzite (Quartz arenite)

Identification: With a bare minimum of cement materials, orthoquartzite looks like solidified sand, which it is. It looks even more like lump sugar because the predominance of quartz makes it very pale in colour.

Orthoquartzites, or quartz arenites, are among the most quartz-rich of all rocks, made almost entirely of sand-sized grains of quartz with a bare minimum of cement. Indeed, the definition of an orthoquartzite is a sandstone consisting of over 95 per cent quartz. There are also a few traces of 'heavy' minerals such as zircon, tourmaline and rutile, but these are fairly scanty.

Orthoquartzites form in high-energy environments. They are created where sand is dropped by waves crashing on beaches and by powerful rivers streaming into the sea. The often dramatic cross-bedding and ripple marks in orthoquartzites are a telling reminder of the way these strong currents tugged the sands to and fro. These sands accumulate above the level where the biggest storm waves start, piling up along the shoreline as beaches, dunes, tidal flats, spits and bars. Look closely at any orthoquartzite formation and you can often see the remnant form of these ancient coastal features.

Not all orthoquartzites form in water. Because they are made of essentially dry sand, they can form from sand-seas in deserts, where sands are piled high by desert winds. Water-formed orthoquartzites tend to be white or pale grey, because they are almost pure quartz. These wind-blown, or 'aeolian', orthoquartzites are often stained red or pink by fine powdered iron oxides, which coat the grains.

About a third of all sandstones are orthoquartzites, but their spread in space and time is patchy. Many formed in a surprisingly narrow time band in the Palaeozoic period (570–245 million years ago). It needed a plentiful supply of continental rock to be weathered, and an unusually long and stable period of weathering to provide all the sand to make them (as well as the removal of other impurities). This is why orthoquartzites are found on the stable margins of ancient continental cratons, such as central Australia, the Russian platform and the St Peters Sandstone of central North America. Some hugely thick orthoquartzites began as deposits where continents rifted slowly and moved apart, which were then folded up into mountain ranges. The Clinch sandstones of the Appalachians and the Tapeats of the Rockies are believed to have been formed by this process.

Grey orthoquartzite: Being almost pure quartz, washed by water or scrubbed by the wind over countless years, quartz arenite or orthoquartzite is often a remarkably clean, pale white or grey quartz colour.

Grain size: Over 95% sand-sized grains, 60μm–2mm/2.5–80mil
Texture: Gritty texture like solid sand. Grains well sorted and well rounded.
Structure: Typically occurs in blanket-shaped deposits varying from a few metres to several hundred metres thick. They are usually interbedded with mudstones, limestones and dolomites. They usually display dramatic cross-bedding and ripplemarks reflecting their formation in high-energy environments. Aeolian rocks often show sand-dune shapes.
Colour: Water-formed orthoquartzites are typically white or pale grey. Aeolian orthoquartzites are often stained red, pink or brown by iron oxides.
Composition: Over 95% quartz, with a smearing of feldspar and carbonate cement. Also chert and metaquartzite, zircon, tourmaline and rutile.
Formation: Some form in high-energy marine environments such as beaches and spits around the edges of stable cratons. Some form on the continental shelf between rifting continents. Some are formed from aeolian (wind-blown) sand-seas in deserts, where there is a ready supply of sand.
Notable occurrences: Russian steppes; central Australia; St Peter sandstone, mid-west USA; Chilhowee, Tuscarora and Clinch formations in the Appalachian Mountains; Flathead and Tapeats formations in the Rocky Mountains

Uluru

Australia's Uluru is the world's largest single block of freestanding rock. Towering to over 345m/1,100ft, Uluru looks like a giant boulder poking out of the sand of the Simpson Desert. In fact, the exposed rock is just the very tip of an ancient outcrop of arkose sandstone extending far under the desert. This outcrop formed when an ancient ocean floor called the Amadeus Basin was uplifted some 550 million years ago, initiating a dramatic period of erosion, and the deposition of arkoses. Later crustal movements have tilted these arkoses almost on end, at 80–85 degrees. Finally, Uluru's arkose was buried beneath the sediments of a shallow sea, and has only re-emerged as wind and water stripped these sediments away. Uluru was long known by its European name, Ayers Rock, but it is a sacred site for Aboriginal peoples, and in 1985 the Australian government gave its custodianship back to the Aboriginals and restored its Aboriginal name, Uluru.

Greywacke

Often called dirty sandstone, greywackes are tough, dark sandstones made from large, sharp grains of quartz, feldspar and rock fragments set in a mass of clay and silt. This unusually chaotic mix was often piled up by submarine avalanches or 'turbidity currents' that plunged from the continental shelf into the deep in a huge, churning mass of water and debris. Deposits are often thousands of metres thick and include the fossils of all kinds of deep-water creatures and plants caught up in the maelstrom. Greywackes were the main sandstones formed early in Earth's history, because land masses were so small at the time. Sandstones formed more recently are often better sorted.

Grain size: Mostly sand, mix of sizes from clay to gravel
Texture: Chaotic mix of sand, gravel and silt. Poorly sorted but well graded.
Structure: Graded bedding. Beds folded and deformed. No cross-bedding. Forms sequences with laminated sandstones and shale.
Colour: Grey, green, brown
Composition: Quartz (40–50%), feldspar (40–50%), mica, plus clay and rock
Formation: Deposited by turbidity currents, and in other high-energy environments
Notable occurrences: All fold mountain belts (except where there is lots of limestone), e.g. Wales; Scottish Uplands, Scotland; Cumbria, England; Schiefergebirge, Harz, Germany; Massif Central, France; Caples, Torlesse and Waipapa terrane, New Zealand; Coast Range, California; West Virginia

Identification: The grey colour and chaotic mix of large fragments amid sand and clay make greywacke easy to identify.

Arkose (Feldspathic arenite)

Arkoses look so much like granite it can be hard to tell them apart. Often bedding marks are the only telltale signs that a rock is sedimentary arkose, not granite. This is because arkose is essentially reconstituted granite, with the same basic ingredients: quartz, feldspar and mica. It is the rock that forms when granite breaks down under particular conditions. What makes it different from other sandstones is that it contains feldspar. Under normal conditions, feldspar is weathered to clay, leaving just clay and quartz. Yet in arkose, feldspar is preserved. It was once thought this meant arkoses could form only in desert environments where there was too little moisture to destroy the feldspar. The Torridonian sandstones of north-west Scotland formed liked this. Now geologists know that feldspar may also be preserved if granite is being eroded and uplifted very rapidly. As a result, many arkoses formed as deltas and alluvial fans, where rivers spill out on to the grabens (depressions) created by the rifting of continents. Others occur along volcanic island arcs. So arkoses are linked with extremes in Earth's past – either extreme climates, or dramatic tectonic movements and high relief.

Identification: Arkose can look very like granite, with the same pinkish colour and the same assemblage of coarse-grained quartz, feldspar and mica minerals. The telltale signs are usually the shape of the formation, and evidence of bedding and layering.

Grain size: Mostly sand-sized grains, at least 1–2mm/ 40–80mil across
Texture: Grains not as well sorted or rounded as ortho-quartzite, except desert arkoses
Structure: In fan-shaped deposits a few metres deep. Less cross-bedding and ripple-marks than orthoquartzites. Aeolian rocks may show sand-dune shapes
Colour: White, grey or pink reflecting feldspar content
Composition: Quartz (40–50%), feldspar (40–50%), mica. In continental arkoses, orthoclase and microcline are the main feldspars; in island arc arkoses, plagioclase dominates
Formation: As deltas and in river bars in areas of high relief and aeolian (wind-blown) deposits in deserts
Notable occurrences: Torridon, Scotland; Pennines, England; France; Czech Rep; Uluru, Australia; Fountain Form., CO Calif; eastern USA

RUDITES

Sandstones, siltstones, mudstones and claystones are all made of small, fairly evenly sized grains.
However, some sedimentary rocks are made from a chaotic jumble of stones of many different sizes.
These jumbled, stone-filled rocks are called rudites, and are divided into two types: conglomerates and
breccias. In conglomerates, the stones are smooth and rounded. In breccias, they are sharp and angular.

Conglomerates

Sometimes called roundstone, conglomerates are basically round stones set in a matrix of finer sand and clay. The stones can be gravel (2–4mm/0.079–0.157in), pebbles (4–64mm/0.157–2.52in), cobbles (64–256mm/2.52–10.08in) and boulders (larger than 256mm/10.08in). Pebbles like these must have been tumbled along beaches or bowled down streams for countless years to round off all the sharp edges – and they had to be tough to survive this battering, so the stones in conglomerate are usually tough materials such as quartz, flint, chert and hard igneous rocks. In time, though, even the hardest stones are reduced to sand and clay. So conglomerates mark an interruption in the slow, steady process of deposition.

There are two kinds of conglomerate – orthoconglomerates and paraconglomerates. Orthoconglomerates are true sedimentary rocks and form where gravel and pebbles are dropped by flash floods in rivers or by storm waves on beaches. The stones in them are quite evenly sized and tightly packed. The spaces in between them are gradually filled up with finer sediment to cement them together, but the rock would be much the same shape with or without it.

Paraconglomerates are formed in one fell swoop and are a jumble of stones of all sizes scattered through a matrix. They are typically formed by landslides, by turbidity currents, and by glaciers – all dramatic events that move material wholesale without any sorting. Take away the matrix and all that is left is a pile of stones. Boulder clay is paraconglomerate.

Conglomerates are widespread, but deposits are usually small and localized. In some, dark pebbles stand out against the light cement like raisins in a pudding, earning them the name puddingstones. The brown puddingstones of Hertfordshire in England and Roxbury in Massachusetts, and the jasper puddingstones of St Joseph Island on Lake Huron in Canada, are all good examples of this.

Puddingstone: With their raisin-like pebbles, the puddingstones of Hertfordshire in England are very striking conglomerates. The white is a cement of quartz and feldspar; the pebbles are flints from the nearby chalk hills.

Petromict conglomerate: The great majority of conglomerates are described as petromict or polymict. This means they contain a wide mix of different stones from a variety of sources, such as basalts, slates and limestones. They are mainly river deposits washed down from areas of high relief and dumped in alluvial fans.

Grain size: Over 2mm/0.079in and can be granules, pebbles, cobbles or boulders

Texture: Orthoconglomerates are mostly gravel-sized grains nearly touching and less than 15% sand and clay matrix. Paraconglomerates are at least 15% matrix and are really sand- or mudstones scattered with pebbles, cobbles and boulders.

Structure: Conglomerates are generally small, poorly stratified deposits with none of the bedding marks of finer sediments

Colour: The colours are as varied as the rocks their stones came from. The stones are often markedly different in colour from the matrix. In jasper puddingstone, red stones are set in a pale matrix, like cherries in a cake.

Composition: The stones can be pure quartz or feldspar from sources such as pegmatites, but usually they are rock fragments – typically harder rocks such as rhyolite, slate and quartzite. The matrix can be silicates, calcites or iron oxides.

Formation: Orthoconglomerates form in fast-moving rivers and in shallow surf. Paraconglomerates are deposited by glaciers, landslides, avalanches and turbidity currents.

Notable occurrences: Hertfordshire, England; Kata Tjuta (Northern Territory), Australia; Huron, Ontario; Keeweenaw, Michigan; Ohio; Indiana; Illinois; Bahamas; Crestone (San Luis Valley), Colorado; Roxbury, Mass; Fairburn, S Dakota; Brooks Range, Alaska; Basin and Range, New Mexico; Van Horn, Texas; Death Valley, California

Breccias

Sometimes called sharpstone, breccia is basically rubble turned into stone. The stones in breccia are jagged, caught up before there was time to round off any rough edges. Unlike conglomerates, breccias can form from almost any rock, soft and hard alike. But they almost always form near to their source and are said to be 'intraformational'. If the stones are washed any further away, they tend to get sorted and so do not form breccias. In mountain areas, breccias often form when screes are cemented together by finer sediment accumulating between the stones. Many breccias are formed rapidly by dramatic events – as when landslides and avalanches come to rest, or when flash floods or storm waves sweep masses of sediment into a beach or bar. Breccias also form when the roofs of limestone caves collapse, burying the floor in rubble. Coral reefs often contain extensive limestone breccias made of fragments broken off the reef. A few breccias are 'extraformational', swept far from their source before consolidating and so have a very mixed composition.

Identification: Breccias are easily recognized by the large, angular stones they contain. It is not so easy to identify what the stones are or where they came from. The best place to start is a comparison with nearby rocks.

Grain size: Over 2mm/0.079in
Texture: Large angular stones in a finer matrix
Structure: Breccias are generally small, poorly stratified deposits
Colour: The colours are as varied as their source rocks
Composition: The stones are usually rock fragments – of almost any rock, including softer rocks such as marble
Formation: Some form in fast-flowing rivers, or on storm beaches. Others form from landslides and avalanches, both on land and under the sea.
Notable occurrences: Thessaly, Greece; Mexico; Vancouver Island, Midway, British Columbia; Platte Co, Wyoming; San Bernardino Co, California; Makinac Island, Michigan; Zopilote, Texas

Landslides

Every now and then a hill or cliff collapses suddenly in a landslide. Some landslides, like Black Ven in Dorset, are triggered as waves undercut the coast. Some are set off by a storm, like the thousands all over New Zealand after Cyclone Bola in 1986. Some are set off by volcanoes and earthquakes, like the 1989 Loma Prieta quake in California. Few events re-shape geology and remake rock material quite so quickly and dramatically. Soft rocks such as clays are very prone to landslides, but tougher rocks can also slide under certain conditions. They tend to fail along existing cracks such as joints. A key factor is often the presence of water, which pushes grains apart and reduces their cohesion. Local rains have caused landslides in the coastal town of Ventura, California (above). Very large rock falls can trap enough air to cushion the fragments, allowing them to travel far and fast. The 1970 Huascaran avalanche in Peru hurtled down the mountainside at a speed of over 320kph/200mph, killing 17,000 people in the towns in its path.

Volcanic, crush and impact breccias

Not all breccias are sedimentary. Volcanic breccias are tuffs that form from fragments blasted out by volcanoes. Crush breccias are formed when rocks are crushed underground by the sheer weight of formations above or by powerful tectonic movements. Some crush breccias are small scale, forming when veins and fissures are squeezed by crustal movements. Others occur on a much larger scale along faults, when the world's tectonic plates crunch past each other, or when layers of rock are folded in mountain building. Meteorite impacts create yet another kind of breccia when the huge force of an impact smashes crustal rocks to bits.

Grain size: Over 2mm/0.079in
Texture: Large angular stones in a finer matrix
Structure: Small, poorly stratified deposits
Colour: The colours are as varied as their source rocks
Composition: The stones are usually rock fragments
Formation: Volcanic breccias form from pyroclasts. Crush breccias form underground when rocks are crushed by crustal movement. Impact breccias form from rocks smashed by meteorite impacts.
Notable occurrences:
Volcanic breccias: Arizona; New Mexico.
Crush breccias: Highlands, Scotland; Alps, Switzerland; Appalachian Mountains.
Impact breccias: Haughton Impact Crater (Devon Island), Nunavut.

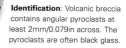

Identification: Volcanic breccia contains angular pyroclasts at least 2mm/0.079in across. The pyroclasts are often black glass.

BIOCHEMICAL ROCKS

*Countless creatures are able to extract dissolved chemicals from seawater and use them to make shell
and bone. Some use calcium and carbon to make carbonates. Others use dissolved silica to make
silicates. When these creatures die, the solid material they created turns into sediments, which form
'biochemical' sedimentary rocks such as chert, flint, chalk and diatomaceous earth.*

Bedded chert (biochemical chert)

Chert is made of quartz crystals so fine they can be seen
only under a microscope. It is an incredibly hard rock, yet
when hit with a hammer it cracks almost like glass into
sharp conchoidal fragments – a quality that was much
appreciated by prehistoric people for making
cutting tools. Most beds of chert formed
from the ooze that covers much of the deep
ocean floor even today. The ooze is built
up from the constant rain of plankton
remains such as radiolarians, diatoms and
microscopic sponges called spicules. Once
the ooze is buried it slowly solidifies into
chert. Relatively pure silica-rich oozes are
known as radiolarian or diatomaceous oozes,
depending on which microscopic organism is
dominant. Slightly less pure oozes are known
as sarls and smarls. Each forms a particular
kind of chert. Ocean bed ooze chert is the top
layer, above serpentines and basalts, in ophiolite
sequences – segments of the sea floor thrown up
on to dry land.

Identification: Chert is easy to
recognize by its very fine-grained,
almost glassy texture, and its
tendency to break into sharp,
conchoidal fragments when
hit with a hammer.

Grain size: The crystals are
cryptocrystalline (too small to
be seen with the naked eye)
Texture: Almost glassy, with
conchoidal fracture
Structure: Biochemical cherts
form in thin layers
1–10cm/0.4–3.9in thick.
Typically massive or finely
laminated (reflecting seasonal
currents), but can show
cross-bedding and scour
marks from turbidity currents.
Colour: Black, white, red,
brown, green, grey,
depending on impurities
Composition: Mostly
pure quartz
Formation: Forms when sea
floor ooze solidifies
Notable occurrences:
Aberdeenshire, Scotland;
Peaks, England; Bavaria, Harz,
Schiefergebirge, Germany;
Bohema, Czech Republic; La
Salle County, Illinois; Marion
Co, Arkansas; Ozarks,
Missouri; Minnesota

Flint (replacement chert)

Identification: Flint nodules look
like white, knobbly pebbles on
the outside, but once broken they
look like black or treacle-toffee
coloured glass – though they are
much harder and break with very
sharp edges.

Not all chert is biochemical in origin. Some is simply
chemical. In other words, the silica is formed without any
organisms, as calcite crystals in limestone are replaced. The
best known of these replacement cherts are flints. Flints are
nodules of black or toffee-coloured chert that form in
limestones, especially the Cretaceous chalks of southern
England and northern France. Any chert that
is black may also be called
flint. Both kinds of flint
were widely used by
prehistoric peoples for
making tools, and also
for striking sparks to
make fire. Chert
formed by
replacement can also
occur as a fine powder
scattered throughout
limestone, and it is also very occasionally found
as the cement in sandstones.

Grain size: The crystals are
cryptocrystalline, which
means they are too small to
be seen with the naked eye
Texture: Almost glassy, with
conchoidal fracture
Structure: Nodules.
Sometimes flint forms around
a network of burrows like
those of *Thalassinoides*
(a branching burrow with Y- or
T-shaped branches), so flint
takes this shape.
Colour: Usually black
Composition: Nearly
pure quartz
Formation: Flint forms from
the solidification of sea
floor ooze
Notable occurrences: North
Yorkshire Moors, North and
South Downs, England;
Rugen, Germany; Mon,
Denmark; Flint Ridge, Ohio

Chalk

Red chalk: Chalk may often be stained red by iron oxides.

Chalk is a white rock of almost pure calcite found in Europe and North America. About 100 million years ago in the Cretaceous period, large lowland areas of these continents were covered with tropical seas. Countless tiny floating algae left plate-like remains called coccoliths across the sea bed some 90–600m/300–2,000ft down, along with the shells of almost equally tiny organisms such as foraminifera. These algal plates and shell fragments turned quickly to almost pure white calcite. The sea bed remained undisturbed for a long time, and layer upon layer of these micro-organisms, along with the occasional larger shells such as ammonites, built up into thick layers of chalk, famously exposed in England's White Cliffs of Dover. Chalk is much softer than other limestones, and the vast beds that once covered most of north-west Europe have been stripped away, leaving bands of rounded hills. Chalks are porous rocks, though not very permeable, and these hills are marked both by dry valleys or bournes formed in wetter times, and also combes created by masses of crumbled rock flowing downhill during colder times.

Identification: Chalk's white colour is unmistakable. It looks like a fine powder, but the coccolith plates and foraminifera shells are clearly visible under a powerful microscope.

Grain size: Very fine-grained like mudstone
Texture: Powdery grains
Structure: Well stratified, with layers often shown up by beds of clays, shell layers and flint nodules. Often has burrow patterns. Occasional layers of crusted material called hardgrounds or Chalk Rock
Colour: White, occasionally red
Composition: Pure calcite
Formation: Forms from the remains of marine algae and microscopic shells
Notable occurrences: North Yorkshire Moors, Downs, Chiltern Hills, England; Champagne, France; Rugen, Germany; Mon, Denmark; South Dakota to Texas to Alabama

---

Stone axes
Flints gave our human ancestors their first tools. Chipped to give a sharp edge, they made it possible to cut through tough hides to get at meat, or, later, to cut hide to make clothes and plants to make tools and shelters. The first stone toolmaker was *Homo habilis*, who appeared about 2.3 million years ago, but it was *Homo erectus* (1.8 million years ago) who made the first crafted stone hand axes. Named Acheulian axes after the French village where they were first found, these axes had two cutting edges and a round end for holding. Axes like these were widely used for over a million years, until half a million years ago a technique for creating a long narrow blade was devised. About 50,000 years ago, modern humans, *Homo sapiens,* made another key breakthrough in blade technology, creating stone knives. Getting a good edge from a flint stone, called knapping, was a tremendously skilled job, and there is evidence that factories were set up where the best knappers would work.

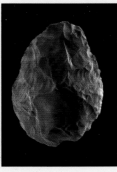

Diatomaceous earth

Diatoms are among the most abundant of all microscopic marine algae. When they sink to the bottom, their minute shells collect in the ooze and eventually turn to what is called diatomaceous earth. When this occurs in a more compact form as a soft, very light, porous, chalky rock it is called diatomite or kieselguhr. Miners sometimes call it white dirt because in bright sunlight it can look like fresh snow. Diatomaceous earth's remarkable purity and fine grain makes it a perfect filtration material, as well as a filler for paper, paint and ceramics. When sugars and syrups are clarified, diatomaceous earth is usually the filter. It is also used as a mild abrasive in toothpastes and polishes.

Grain size: Very fine-grained like mudstone
Texture: Powdery grains. Diatom shells can be seen under a powerful microscope
Structure: Well stratified, with layers often shown up by beds of clays
Colour: White, yellow, greenish grey, sometimes almost black
Composition: Silica shells of diatoms
Formation: Diatomaceous earth forms from the remains of marine algae and microscopic shells
Notable occurrences: Denmark; Lüneburger, Saxony, Halle, Germany; France; Central Italy; Russia; Algeria; Nevada; Oregon; Washington; Santa Barbara, California

Identification: Diatomite looks a little like chalk but is so light it almost floats on water like pumice.

LIMESTONES (CARBONATE ROCKS)

Made up of at least half calcite (or the similar aragonite), limestones are distinctive whitish, grey or cream rocks. They are the third most abundant sedimentary rocks on Earth, after mudrocks and sandstones, and extend over vast areas of continents and continental shelves, dominating many mountain chains. Limestone 'karst' landscapes can often be very dramatic, with their caverns and gorges.

Limestone

Coral limestone: Few rocks are richer in fossils than limestones. Very often you can see perfectly preserved remains of sea creatures that swam and crawled in tropical seas long ago, or, like the coral polyps preserved in this rock, simply sat on the sea floor and waited for a meal to pass.

Limestones are a striking testament to the sheer profusion of life on Earth, especially in the sea. They are almost entirely the work of living things. Huge beds of limestone thousands of metres thick may be the accumulated remains of countless sea creatures piled up on the sea bed over millions of years, then slowly changed into rock as their chemistry alters. This accumulation is going on today, notably in places such as the Bahamas, and these remains too will in time turn to rock.

Living things contribute to the creation of limestone in two ways. Sometimes they contribute their 'skeletal' remains, their hard shells and bones, to the rock. Alternatively, like plankton and algae, they change the chemistry of the sea, and encourage the deposit of calcite. The key chemicals in limestones are carbonates – and in particular calcium carbonate in the form of calcite or aragonite. Carbonate sediments may be rich in either calcite or aragonite, but ancient limestones are almost always calcite-rich because aragonite alters over time to calcite.

Limestones form in many places – soils on old rocks, river flood plains, lakes – but most are the creation of shallow, clear tropical waters. Here there is not only an abundance of sea life, but the evaporation of such warm waters boosts the precipitation of calcium carbonates. This does not mean limestones are found only in the tropics, however. The continents have shifted so much through the ages that many places now nearer the Arctic were once in the tropics. During the Carboniferous period around 300 million years ago, much of what is now North America and Europe lay in the tropics, and was inundated by vast tropical seas. Huge beds of limestone, now visible in places such as Texas and the English Pennines, are the legacy of this time. In England, such limestones are called Carboniferous limestones.

Reef limestone: Reef limestones are the work of corals, those remarkable sea creatures still building up huge colonies like Australia's Great Barrier Reef. Reef limestones are made partly from their skeletons, built up over the ages, and partly from sediment trapped and bound by mats of microbes living on the reef. Unlike some other limestones, reef limestones contain no visible skeletal remains. They are also harder than other limestones, and are often left protruding as small hills after softer surrounding limestones has been weathered away.

Grain size: Varies from clay-sized to gravel-sized
Texture: Highly variable, from very fine-grained, porcelain-like look to aggregate of large fossils
Structure: Most limestones show the same range of structures as sandstones and mudrocks. Beds often include reef limestones, the fossils of coral reefs. Reef limestones show little bedding, although they preserve the growth pattern of corals and cavities filled by carbonate debris and cement. Patch reefs or 'bioherms' are oval lumps left by small round coral colonies. 'Biostromes' are large long limestone formations left by barrier reefs.
Colour: White, grey, cream plus red, brown, black
Composition:
Skeletal remains: Algae and microbes (coccoliths and stromatolites); Foraminifera; Corals; Sponges; Byrzoans; Brachiopods; Molluscs; Echinoderms; Arthropods. Carbonate grains (overleaf): Ooids and pisoids; Peloids; Aggregates; Intraclasts. Lime mud: Bone, teeth and scale debris (phosphates); Wood, pollen and kerogen (carbonates); Cement (calcite, aragonite, dolomite).
Formation: Forms as carbonates form, mainly on the sea floor, either from the skeletal remains of sea creatures or by the precipitation of calcite
Notable occurrences: Burren, Ireland; Pennines, Cotswolds, England; Slovenia; Italy; Swartberg, South Africa; Ratnapura, Sri Lanka; Laos; Thailand; Guilin, China; Victoria, Australia; Paparoa, New Zealand; New Mexico; Kentucky; Texas; South Dakota; Indiana; Onondaga, New York

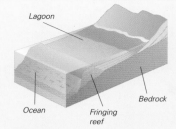

Coral reefs
Coral reefs are one of the wonders of tropical seas, teeming with an astonishing variety of sea creatures. The reefs themselves are made up from tiny sea anemone-like animals called polyps, which stay all their lives fixed in one place attached to a rock or to dead polyps. They take dissolved calcium carbonate from seawater and turn it into the mineral aragonite to build the cup-shaped skeleton or corallite in which they live. The skeleton becomes hard coral when they die. Coral reefs are made from millions of polyps and their skeletons, and can stretch for thousands of kilometres. Fringing coral reefs grow at a particular depth along the shoreline. Barrier reefs form a little way offshore. Coral atolls form around the edge of an island volcano. As the volcano sinks or the sea level rises, so the coral grows up and eventually leaves just a ring or atoll. Corals have been around since Cambrian times, and their reefs and fossils are abundant in limestone rocks of all ages since then.

Fossiliferous limestone: Bryozoan limestone

Like crinoids, bryozoans were sea creatures that lived in such vast numbers in the tropical oceans of the past that their remains have gone on to make a specific and abundant kind of limestone, bryozoan limestone. Over 15,000 species of bryozoan have been identified, of which 3,500 are alive today, living in many ocean shallows such as the western Pacific. They live in colonies of hundreds of animals or zooids, each secreting a short tube of lime to enclose its soft parts. A ring of about ten tentacles snakes out from the end of the tube to guide food into the animal's mouth.

Bryozoan colonies look so like lace, they are also known as sea lace.

Grain size: Sand-sized grains with fossil remnants
Texture: Highly variable, with sand-sized grains and partial and complete fossils
Structure: Marked cross-bedding and ripple-marks. Layering from repeated cycles of sedimentation. Often broken into massive blocks divided by vertical joints and horizontal bedding planes
Colour: White, cream
Composition: Calcite
Formation: From bryozoans in shallow tropical seas
Notable occurrences: North Wales; Norfolk, England; Southern Sweden; Stevns Klint (Zealand), Denmark; Moravia, Czech Republic; Torquay and Geelong (Victoria), St Vincent (SA), off Tasmania, Australia; off Otago, Oamaru (South Island), New Zealand; Biscayne, Florida; Indiana

Identification: Bryozoan limestone is identified from the lace-like colonies of bryozoans. Individual animals are tubes about 2mm/0.08in long.

Fossiliferous limestone: Crinoidal limestone

Many limestones consist largely of recognizable fossils of ancient sea creatures. Among the most widespread of these 'fossiliferous' limestones are crinoidal limestones. Sometimes called sea lilies, the crinoids of the past were animals that looked like long-stemmed flowers, with a central 'cup' containing the soft parts of the animal, numerous branching 'arms' and a stem up to 30m/98ft 5in long, which attached the animal to the ocean floor. In the Carboniferous period in particular, crinoid flowers grew in such extraordinary profusion that they created vast 'meadows' on the sea floor. When the animals died, ocean currents broke up most of their skeletal plates into sand-sized grains and rolled them together until they were cemented by calcite into thick deposits of limestone. Dramatic cross-bedding in these rocks bears witness to the shallowness of the seas in which the crinoids grew, and the power of the waves and currents that broke up their remains. Whole fossils are rare. The volume of crinoidal limestones around the world is staggering, and incorporates the remains of a huge number of crinoids. There are estimated to be at least 60,000km³/14,400 cu miles of crinoid remains in the Mission Canyon-Livingstone formation in the Rockies alone.

Crinoid fossil

Identification: Crinoidal limestone is stuffed full of the fossils of crinoids. Although very few remain intact, there are usually enough of the cups, arms and stems surviving to be recognizable.

Grain size: Sand-sized grains with fossil remnants
Texture: Highly variable, with sand-sized grains and partial and complete fossils
Structure: Marked cross-bedding and ripple-marks. Layering from repeated cycles of sedimentation. Often broken into massive blocks divided by vertical joints and horizontal bedding planes.
Colour: White, grey
Composition: Calcite
Formation: From meadows of crinoids in shallow tropical seas
Notable occurrences: North Wales; Derbyshire, Durham, Somerset, England; Austria; Nile Valley, Egypt; Timencaline Wells, Libya; Nepal; Namoi, Bingleburra, Australia; Mission Canyon-Livingston (Rocky Mts), Canada–USA; Leadville, Colorado; Redwall, Arizona; Burlington, Iowa to Arkansas

OOLITHS AND DOLOSTONES

There is a huge variety of carbonate rocks. While some limestones are largely fossiliferous or shelly – made largely of fragments of shell and bone – others consist of grains formed by the precipitation of calcite and aragonite from carbonate-rich sea-water. Like limestones, dolostones are carbonate rocks, but they are made of magnesium carbonate instead of calcium carbonate.

Grain limestone (Oolitic and Pisolitic limestone)

Limestones show the same range of grain sizes and textures as sandstones and mudrocks. Indeed, some geologists describe them using the same terms (lutites, arenites and rudites), adding 'calci-' or 'calca' to show they are limestones. So calcilutites are limestone muds, calcarenites lime sands and calcirudites gravels. Many calcarenites contain only a few shellfish remains. Instead, they are made mostly of calcite or aragonite grains precipitated out of water. Calcite grains can be washed into deposits just like sand and mud grains, but most form in situ. Ooliths or ooids are tiny balls made as layers of calcite build up on clay grains, kept round as they are rolled by underwater currents. Pisoids are gravel-sized balls that form in the same way. They look similar to grains called oncoids, but oncoids are actually created by microbes. Peloids are oval grains that usually started life as pellets of snail and shellfish faeces and were then altered to micrite (fine-grained calcite). Intraclasts are bits of broken calcite sediment. Limestones can be classified according to the dominant type of grain, as shown in the table below. They can also be classified by their texture (see table opposite).

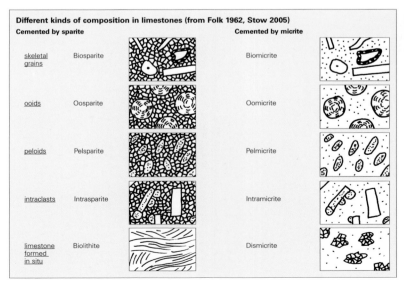

Oolitic limestone: Also known as roestone because it looks like fish roe, oolitic limestone is made of sand-sized grains called ooliths. Ooliths form in shallow, carbonate-rich tropical waters such as those around the Bahamas today, starting as aragonite and later changing to calcite. Wherever oolitic limestones appear, they are reminders that conditions were once like this.

Grain size: Ooids (0.2–0.5mm/7.87–19.69mil); pisoids and oncoids (over 2mm/78.8mil); peloids (over 1mm/39.4mil); intraclasts (1mm–20mm/0.04in–0.8in)
Texture: (see table opposite)
Structure: Grain limestones show the same range of structures as sandstones and mudrocks
   Colour: White, grey, cream plus red, brown, and black
Composition: Carbonate grains with lime mud cement (sparite and micrite)
Formation: Forms by the precipitation of aragonite in shallow, carbonate-rich tropical waters
Notable occurrences: Dorset, Cotswolds, England; Luxembourg; Harz, Thuringia, Germany; Ukraine; Kertsch, Russia; Caucasus, Georgia; Newfoundland; Texas; Alabama

Different kinds of composition in limestones (from Folk 1962, Stow 2005)

	Cemented by sparite			Cemented by micrite	
skeletal grains	Biosparite			Biomicrite	
ooids	Oosparite			Oomicrite	
peloids	Pelsparite			Pelmicrite	
intraclasts	Intrasparite			Intramicrite	
limestone formed in situ	Biolithite			Dismicrite	

Pisolitic limestone: Oolitic limestone is made from sand-sized grains 0.2–0.5mm/7.07–19.69mil in diameter; pisolitic limestone is made from larger, pea-sized grains at least 2mm/78.8mil across.

Dolostone (Dolomite limestone)

Ever since they were first identified in 1791 by Frenchman Deodat de Dolomieu in the Italian Dolomite mountains named after him, dolostone or dolomite limestone has intrigued geologists. While ordinary limestones are made of calcite or aragonite, dolostones are at least half made of the magnesium carbonate mineral dolomite. Until the 1960s, when it was found forming along the shore in the Arabian Gulf and in the Bahamas and Florida, no one had seen it actually forming directly from seawater. It seemed as if all dolostone formed by the chemical alteration of calcite in limestones by magnesium-rich solutions, a process called dolomitization. This is probably how most dolostones did form, but the process is now better understood. It seems to involve salty brines formed as evaporation concentrates seawater in tropical lagoons. These magnesium-rich briny waters sink seawards through limestones, slowly turning their calcite to dolomite. This process was more prevalent in the past, and most dolostones are Precambrian in origin (at least half a billion years old).

Identification: Dolomite is a much tougher rock than limestone and has a sugary white crystalline look. Recrystallization destroys fossils, so there are no visible organic remains.

Grain size: Varied – some forms are microcrystalline; others are sand-sized
Texture: Dense sugary texture
Structure: Stands out from ordinary limestone in rib-like beds because it is so tough. Coarse crystal dolostone shows the same structures as other limestones; fine crystalline dolostone does not.
Colour: White, grey, cream but weathers pink or brown
Composition: At least half dolomite
Formation: Thought to form when calcite in limestone is dolomitized (recrystallized and turned to dolomite)
Notable occurrences: Central England; Swabian and Franconian Jura, Rhineland, Germany; Dachstein, Austria; Dolomites, Italy; Niagara, Ontario; Arkansas; Iowa; Ohio; Kentucky

Karst scenery
Limestone may have formed in water and yet the calcite it is made of is also quite easily dissolved by water that is slightly acidic. Rain and groundwater take up carbon dioxide from the air and soil, turning them into weak carbonic acid. Wherever limestone is near the surface, this acidic water seeps into cracks and begins to dissolve the rock. After thousands of years, huge cavities can be etched out often creating spectacular scenery, known as Karst after the Kras plateau in Slovenia, one of the many places where such scenery is found. Underground, huge potholes and caverns with stalactites and stalagmites are created. Above ground, cracks around blocks of rock on surfaces are etched out to create striking limestone pavements. Often cavern roofs collapse or potholes grow and merge to create deep gorges. Eventually, so much rock will be dissolved away that only distinctive, towerlike pillars are left, such as these in the famous Guilin Hills of China (above).

Different kinds of deposit texture in limestones

Original components not bound together

Mudstone (mud-supported, less than 10% grains)

Wackestone (mud-supported, more than 10% grains)

Packstone (grain-supported)

Grainstone (lacks mud and is grain-supported)

Original components bound together

Boundstone

No recognizable deposition texture

Crystalline

Original components not organically bound

Floatstone (matrix-supported, less than 10% sand-sized grains)

Rudstone (sand-supported, less than 10% sand-sized grains)

Original components organically bound

Bafflestone (organisms act as baffles)

Bindstone (organisms encrust and bind)

Framestone (organisms build a rigid framework)

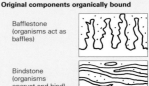

This classification was devised by Dunham in 1962, then modified by Embry and Klovan in 1971 and Stow in 2005

CHEMICAL ROCKS

Chemical sedimentary rocks form neither from debris nor with the aid of living things, but entirely chemically as minerals precipitated out of water solutions. Many are left behind as solid 'evaporites' when the solution evaporates. Yet precipitation can occur whenever a solution becomes saturated and can no longer retain the minerals dissolved in it, forming rocks such as tufas, travertines and dripstones.

Tufa

Tufas are calcite deposits that build up around the rim of calcite-rich springs, rather like the limescale that builds up in baths and taps in areas of hard water. Tufa often builds towers underwater where springs bubble up into lakes or under the sea. If lake levels drop, these towers may be exposed, as in California's famous Mono Lake. Although tufas are chemical rocks, algae and other plant material does play a part in their formation. Tufa is always precipitated on to some surface or other, and quite often the surface is algae or plants. Indeed, algae actively spurs tufa to precipitate, forming algal mats or mounds called stromatolites made of tufa bound together by filaments of algae. Often the algae rots away leaving a sponge-like rock called a sinter. Tufa is sometimes called calcareous sinter to distinguish from siliceous sinter, a sinter that forms by the precipitation of opaline silica. Because it is so full of holes, tufa is light and easy to cut, which is why the Romans used it to line the Aqua Appia, the underground aqueduct they built in 312BC to supply the city of Rome with water.

Identification: Full of holes like a sponge, and quite light and soft, tufa is easy to recognize. It is usually white or a buff colour, but iron oxides can turn it red or yellow. Deposits are usually quite thin.

Grain size: Powdery
Texture: Compact to earthy and friable
Structure: Tufa is spongy and full of holes. Structures take the form of the places that they formed. Towers form around underwater springs. Algal colonies often form mounds.
Colour: White, buff, yellow, red
Composition: Calcium carbonate in the form of calcite, or occasionally aragonite
Formation: By precipitation from calcium-rich waters, typically in streams, around springs or algal mounds.
Notable occurrences: Glen Avon, Scotland; Ikka Fjord, Greenland; Great Rift Valley, Kenya; Kimberley, Western Australia; Mono Lake, Mojave Desert, California

Travertine

Identification: Denser and more compact than tufa, with fewer holes, travertine looks a little like tofu, and is usually an attractive pale honey colour.

Tufa forms mainly around cool springs, typically when plants take carbon dioxide from the water and make less available to combine with calcium. Around hot springs, calcite is precipitated when hot water loses carbon dioxide as it cools. This leaves dense, hard crusts, such as those around Mammoth Hot Springs in Yellowstone Park, Wyoming. The terms tufa and travertine are sometimes used interchangeably, but geologists usually call the dense variety travertine, and the spongy variety tufa. Travertine is a pale honey colour, often with delicate banding. Many sculptors have used it as an easier-to-carve alternative to marble, and it is also cut into slabs and made into polished floors. The most famous travertine is Roman travertine, which gave the rock its name.

Grain size: Powdery
Texture: Compact to earthy and friable
Structure: Much denser than tufa, with only a few holes. Often banded.
Colour: Honey, red, brown
Composition: Calcite, or occasionally aragonite
Formation: By precipitation from calcium-rich waters around hot springs or in caves (see Dripstone)
Notable occurrences: Bohemia, Czech Republic; Aniene River, Italy; Pammukale, Turkey; Algeria; Thebes, Egypt; San Luis, Argentina; Baja, Vera Cruz, Mexico; Yavapai Co, Arizona; Yellowstone, Wyoming; Jemez, New Mexico; San Luis Obispo Co, California

Evaporite

In arid conditions, salty water may evaporate to leave dissolved minerals as evaporite deposits. Some evaporites form when desert salt lakes dry up. They form on a larger scale when seawater evaporates in lagoons, coastal shallows and salt flats called sabkhas. The scale of some seawater evaporations is staggering. Evaporites are highly soluble, so they are rare on the surface, but there are ancient evaporites thousands of metres thick dating from the Cambrian, Permian, Triassic and Miocene periods. A depth of 1,000m/3,280ft of seawater needs to steam off to form each 15m/49ft of deposit. So, to build up these massive beds, coastal flats must have been flooded by the sea again and again over vast time spans. There are a large number of minerals dissolved in seawater, but only a few are abundant, and they always tend to be deposited in the same sequence, creating a bull's eye pattern of deposits. The sequence starts with the least soluble, dolomite, then goes through gypsum, anhydrite and halite (rock salt) to finish with the most soluble, potassium and magnesium salts called bitterns. The evaporites formed in salt lakes inland are typically dominated by halite, gypsum and anhydrite, but there are also many more minor salts.

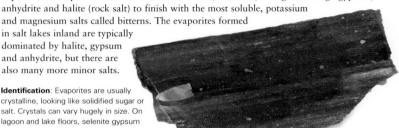

Identification: Evaporites are usually crystalline, looking like solidified sugar or salt. Crystals can vary hugely in size. On lagoon and lake floors, selenite gypsum crystals can grow up to 1m/3ft 3in.

Grain size: Crystal size varies
Texture: Coarse/fine, earthy, friable, massive, sugary
Structure: Lagoon deposits: cracked. Deep water deposits: laminated. Nodules of anhydrite form in sabkha gypsum, often leaving just a 'chicken wire' mesh of gypsum between them.
Colour: White, pink, red
Composition: Dolomite, gypsum, anhydrite, halite or bittern
Formation: Evaporation of salty waters in coastal salt flats, lagoons and salt lakes
Notable occurrences: *Currently forming*: Caspian Sea, Georgia; Persian Gulf; Dead Sea; Great Salt L, Utah. *Ancient formations*: Lakes: Green River, Wyoming. Sabkha and shallow shelf: northern Europe; Elk Pt, British Columbia; Salina, Michigan; Williston, Montana; Delaware, Texas. Deep sea: Mediterranean.

The great Mediterranean salt pan
In 1970, a team drilling in the Mediterranean Sea made an extraordinary discovery. There beneath the sea were the thickest evaporites ever found, many thousands of metres thick. It transpires that at the end of the Miocene epoch, about five million years ago, the movement of the continents greatly narrowed the Straits of Gibraltar. A brief ice age triggered a global drop in sea level, and suddenly the Atlantic stopped flowing into the Mediterranean to keep it topped up. Within a few thousand years, the entire Mediterranean – some 2.5 million km³ (599,782 cubic miles) of water – had evaporated to create one gigantic salt basin, like the Dead Sea but 4,000m/13,123ft deep! Buried in the sediments under the Nile in Africa is a great canyon 2,500m/8,202ft deep dating from this time – indicating that the Mediterranean had dried up entirely. This whole episode is known as the Messinian Event because the best known deposits from the time are under the port of Messina, Sicily (situated just south of the peninsula enclosing the Strait of Messina, shown in the satellite image above).

Dripstone and flowstone

Although most travertines form around hot springs, the most spectacular and beautiful are often those that form in limestone caverns. Here calcite-rich waters dripping from the ceiling create deposits called dripstones. Dripstones can build up in all kinds of fantastic formations, such as stalactites hanging from the ceiling and stalagmites projecting from the floor (see Calcites and Dolomite). Sliced across, these dripstones usually reveal how they were built up in layers, like the layers of an onion, in darker and lighter bands. Cavern walls and floors continually wet with running water may be coated in sheets of travertine called flowstone.

Grain size: Powdery
Texture: Compact to earthy and friable
Structure: Dense, compact. Stalactites and stalagmites show 'growth rings' caused by variations in precipitation.
Colour: Honey-coloured, red, brown
Composition: Calcium carbonate in the form of calcite, or occasionally aragonite
Formation: By precipitation from dripping and flowing calcium-rich groundwaters in limestone rock
Notable occurrences: Kent's Cavern (Devon), England; Skocjan, Slovenia; Aggtelek, Hungary; Sorek, Israel; Reed Flute (Guilin), China; Philippines; Carlsbad, New Mexico; Mammoth Cave, Kentucky; Luray, Virginia

Identification: Dripstones are often markedly layered, revealing variations in the seasonal flow of water down through the limestone.

ORGANIC ROCKS

Coal is a very unusual sedimentary rock. Not only does it burn, which makes it a very useful fuel, but it is also almost entirely organic. It is made not from grains of minerals, like other sediments, but from the remains of plants that grew in tropical swamps hundreds of millions of years ago, transformed into solid black or brown carbon as they were buried.

Coal formation

Most of the coal resources in North America, Europe and northern Asia formed in the Carboniferous and early Permian periods, around 300 million years ago. At this time, these continents lay mostly in the tropics, and vast areas were covered in steamy swamps where giant club mosses and tree ferns grew in profusion. Waters moved through these swamps only sluggishly, so when plants died, their remains piled up on the swamp floor and rotted only slowly in the stagnant, poorly oxygenated water. Microbes began to turn the remains into peat. Peat is about half carbon and is a useful, if smoky, fuel when dried, but to transform peat into coal, it must be deeply buried, to a depth of at least 4km/2.5 miles.

Over millions of years, the peats from the Carboniferous swamp were buried under layers of accumulating sediment until they were not only squeezed completely dry, but began to cook in the heat of the Earth's interior. Cooking did not only destroy plant fibre – it drove out hydrogen, nitrogen and sulphur as gases, and gradually transformed the carbon compounds in the plants to pure carbon. The deeper and longer they were buried and the hotter they got, the more plants turned to carbon. Peat is quite soft and brown and only about 60 per cent carbon; anthracite, the deepest and oldest kind of coal, is hard, black and over 95 per cent carbon. In between come intermediate 'ranks' of coal – brown coal, or lignite (73 per cent carbon), and dull black bituminous coal (85 per cent carbon).

Most black coal dates from the Carboniferous and early Permian. The world's largest resources of these coals are in Russia and the Ukraine, which has almost half the world's entire coal reserves, but there are also huge black coal beds in the USA. Besides black coals, there are brown coals formed more recently, especially in the Tertiary, 1.6 to 64 million years ago. Although less rich in carbon, these coals are widespread, found in China, North America, especially Alaska, as well as southern France, central Europe, Japan and Indonesia.

Peat: All coal may have begun as peat. If so, the peat of old must have formed in tropical swamps. Peat today forms only in bogs in cool places. Here decomposition of plant material is so slow that thick layers can build up before the plants totally rot. Microbes do get to work, however, converting at least half of the plant material to carbon as it becomes compacted.

Lignite: Once peat is buried deeply, the process of 'coalification' begins. Microbial activity ceases, but pressure and heat begin to turn more of the plant remains to carbon. Lignite or brown coal is the first stage. It crumbles when exposed to the air and has a texture like woody peat. Most lignite is more recent than black coal, dating from the Tertiary. It is found nearer the surface than black coal, but is much less carbon-rich and burns with less heat and more smoke.

Grain size: Fine-grained, similar to mudrocks
Texture: Varies with coal rank. The lower ranked coals contain many only partly altered plant remains; the highest ranked coals contain very few. Peat contains un-decomposed plants.
Structure: Coal occurs as beds or seams interlayered with other sedimentary rocks, often with a thin layer of carbon-affected material called seat-earth beneath. Seams are generally only a few metres thick, but can be several hundred metres. Coals form in stagnant environments so do not show any cross-bedding, but humic coals (see Formation below) contain bands from 1–10mm/ 0.039–0.39in thick. Each seam has its own banding profile, which can help to identify it almost like a fingerprint.
Colour: Brown, black
Formation: Most coals are 'humic' and form when plant material piles up in situ in tropical coastal swamps and is then buried deep and converted by heat to carbon. Rarer 'sapropelic' coal forms when plant debris, spores, pollen and algae pile up far from their original source.
Notable occurrences: Some of the world's biggest coal reserves are in the heart of Siberia in Russia, in Kazakhstan and the Ukraine. There are also major reserves of black coal in northern Europe, the Damodar Valley in India, and Appalachians and Midwest of North America. Germany and China have huge resources of brown coal. Pennsylvania is famous for its anthracite deposits.

Coal types and components

Plants are made from a wide range of different components including massive, hard trunks, soft leaves and tiny seeds and spores. Once a plant dies and falls into a swamp, oxygen and microbes get to work on each of these components differently. These differences are most significant in the early stages of the coal formation process, when peat forms. Consequently, peats can differ widely in character according to the plant parts involved. Even once peat is buried and coalification proper begins, the variations in plant parts makes a difference to the nature of the coal. The plant components in coal are called 'macerals', and divided into three broad groups: vitrinite, liptinite and inertinite. Vitrinite comes from the woody parts of the plant – trunks, branches, roots. It's tough and shiny and is the major component in a type of coal called vitrain. Liptinite comes from the waxy and resinous parts of the plant – the seeds, spores and sap. It is softer and duller than vitrinite. Mixed with

Bituminous coal:
Bituminous or soft coal is second only in rank to anthracite, with a 75–85 per cent carbon content. It is dark brown to black and banded, and usually made of over 95 per cent vitrinite, which comes from plants' woody parts. This is the most widely used type of coal, but its high sulphur content can contribute to the creation of acid rain when it is burned.

Composition:
Peat: Over 75% water by weight; Solid matter: over 50% carbon, under 50% dry mineral-free volatiles.
Lignite: 33–75% water by weight; Solid matter: 50–60% carbon, under 50% dry mineral-free volatiles.
Sub-bituminous: 10–32% water by weight; Solid matter: 60–75% carbon, 35–42% dry mineral-free volatiles.
Bituminous: Under 10% water by weight; Solid matter: 75–85% carbon, 18–37% dry mineral-free volatiles.
Anthracite: No water; Solid matter: 75–85% carbon, 18–37% dry mineral-free volatiles.

Plant components (macerals):
Vitrinite group (50–90%): Woody tissue – polymers, cellulose, lignin.
Liptinite group (5–15%): Waxy parts of plant – seeds, spores, resins.
Inertinite group (5–40%): Shiny black plant material highly altered during peat formation.

Coal types:
Vitrain: Glassy, brittle, bright bands, conchoidal fracture, dominated by vitrinites.
Clarain: Finely laminated, silky, bright and dull bands, smooth fracture, mix of vitrinite and liptinites.
Durain: Hard, dull, matlike, dull bands, mix of inertinites and liptinites.
Fusain: Soft, powdery, charcoal-like, dirties fingers, mostly inertinite.

Inorganic components:
Detrital quartz, heavy minerals, sulphates, phosphates, pyrite nodules, marcasite, siderite, dolomite, calcite.

vitrinite, it makes a silky, laminated kind of coal called clarain. Inertinite comes from plant material much altered by oxidation during peat formation, or from parts affected by fungus. Mixed with liptinite, it makes a dull hard coal called durain. By itself it makes the soft, powdery charcoal-like kind of coal called fusain – easy to identify because it leaves your fingers smeared black.

Each coal seam contains varying amounts of these different kinds of coal, but the higher the rank of the coal – the closer to pure carbon anthracite it gets – the more they lose their distinctiveness as a result of the greater degree of 'coalification'.

How coal is mined
The way companies mine coal depends partly on the depth of the seam. With a seam less than 100m/328ft below the surface, the cheapest method is to simply strip off the overlying material with a giant shovel called a dragline. Brown coal tends to occur near the surface, and can often be mined economically by strip mining. The best bituminous and anthracite coal typically lies in narrow layers called seams, far below ground. To get at the coal, mining companies have to sink deep shafts to reach the seam. The Ashton pit in northern England plunged almost 1,000m/3,280ft. With the shaft dug, they then created a maze of horizontal or gently sloping tunnels to get into the seam and extract the coal. The surface of the exposed seam is called the coal face. Mining operations can be hazardous, with the constant danger of roof-falls or of explosions as methane gas forms from the coal. Miners can also suffer lung damage by inhaling coal-dust.

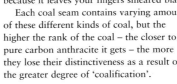

Anthracite: Anthracite, or hard coal, is the highest ranked of all the coals – shiny black and over 95 per cent carbon. Temperatures in the ground have to reach over 200°C/392°F to turn bituminous coal to anthracite. It is the rarest, and usually most ancient, of coals but it has a very high energy content and burns almost without smoke.

METAMORPHIC ROCKS: NON-FOLIATED

Metamorphic rocks are formed from neither melts nor sediments but are created deep underground when other rocks are remade by heat and pressure, sometimes by direct contact with hot magma, sometimes by the tremendous forces present in the Earth's crust. The original rock's minerals are cooked and recrystallized in new forms or even as completely new minerals. Metamorphic rocks are divided into foliated (striped) rocks and non-foliated rocks, which include hornfels, metaquartzite and granofels.

Hornfels

Hornfels is a tough, splintery rock that gets its name from the German for 'horn rock' because broken edges are translucent like horn. Making it involves less stress than other kinds of metamorphism. The original rock or protolith is simply cooked by close contact with an intrusion. The heat is tremendous – typically about 750°C/1,350°F – but the rock is neither crushed, twisted nor pulled. So hornfels is free from foliation. Crystals are fine-grained and point in all directions. Indeed, hornfels can easily look like a volcanic rock. Small structures in the protolith are obliterated during metamorphism. The crystals reform in a tight, interlocking pattern like crazy paving called pfiaster structure, visible under a magnifying glass. Some hornfelses may also be distinctively 'spotted' with porphyroblasts (large crystals like phenocrysts in igneous rock), such as andalusite hornfels and cordierite hornfels.

Hornfels is not just a type of rock, though, but helps identify some of the various 'facies' of metamorphic rock – the particular combinations of minerals formed in different pressure and temperature regimes. Hornfels facies include the hornblende hornfels and the pyroxene hornfels facies. These facies are the array of minerals that form when pressure is low but temperatures are high. The exact composition depends on both the original rock and the temperature, often grading through different minerals the nearer to the intrusion the rock forms and the hotter it gets.

Hornfelses are often divided into three groups according to their protolith: those made from shales and clays; those made from impure limestone; and those made from igneous rocks such as dolerite, basalt and andesite. All these are fine-grained.

Shales and clays form biotite hornfelses specked with black biotite mica, though they also contain feldspar and quartz, and a little tourmaline, graphite and iron oxide. Geologists look in these rocks for the aluminium silicates andalusite, kyanite and sillimanite. Each forms at a particular temperature and pressure, so finding one reveals the conditions in which the rock formed.

Impure limestone hornfelses are tough rocks containing calcium-rich silicates such as diopside, epidote, garnet, sphene, vesuvianite and scapolite, as well as feldspars, pyrites, quartz and actinolite. The igneous hornfelses are, like their protoliths, rich in feldspar with brown hornblende and pale pyroxene, but they also contain streaks and patches of new minerals such as aluminium silicates.

Identification: Plain hornfels like this is very easily confused with basalt and other dark volcanic rocks. Sometimes, though, hornfels is unmistakably 'spotted' with porphyroblasts of minerals such as andalusite, cordierite, garnet or pyroxene.

Striped hornfels: Many hornfels rocks are streaked with crystals of aluminium silicates such as sillimanite and andalusite. These minerals are very characteristic of the hornfels facies, marked by high temperatures and low pressure.

Rock type: Non-foliated, contact metamorphic
Texture: Even, fine-grained; sometimes contains porphyroblasts (large crystals), or poikiloblasts (large crystals enveloping smaller crystals)
Structure: Most small structures are obliterated by metamorphism, though bedding from protolith may be preserved
Colour: Black, bluish, greyish, often speckled with dark porphyroblasts
Composition: The matrix is too fine-grained for individual minerals to be easily distinguished, but tiny flakes of mica can sometimes be seen under a magnifying glass. Square black or red porphyroblasts of andalusite are visible in andalusite hornfels. If these crystals are cross-shaped, they are known as chiastolite, and the rock is called chiastolite hornfels. In cordierite hornfels, the rock is dotted with rice-grain-like porphyroblasts of dark cordierite. In pyroxene hornfels, there are porphyroblasts of pyroxene, andalusite or cordierite. Other common minerals may be garnets, hypersthene and sillimanite.
Protolith: Fine-grained rocks including shales, clays, impure limestones, dolerite, basalt, andesite
Temperature: Very high
Pressure: Low
Notable occurrences: Comrie (Perthshire), Scotland; Cumbria, Dartmoor (Devon), Cornwall, England; Vosges, France; Harz Mountains, Germany; Elba, Italy; Nova Scotia; Sierra Nevada, California

(Meta)quartzite

(Meta)quartzite is a tough, whitish, sugary-looking rock that looks rather like white marble, but is made from quartz, not calcite. Indeed, it is over 90 per cent quartz. It forms mainly from sandstone. Like the sandstone orthoquartzite, it is often simply called quartzite, and the two often grade into each other, depending on how much the original sandstone has been altered by metamorphism. During metamorphism, the quartz grains in sandstone recrystallize, creating new, larger grains. The cement and open pores in sandstone vanish, leaving only tightly interlocking grains. In fact, the quartz grains become effectively welded together so that when the rock breaks, it fractures right across the crystals, rather than breaking around the grains as sandstone does. Most metaquartzite is non-foliated. However, under extreme heat and pressure, it may be flattened or sheared in such a way that the grains are stretched out in a pancake shape, creating foliated metaquartzite.

Identification: (Meta)quartzite is a white rock that looks like marble but is much tougher. Unlike marble, it cannot easily be scratched with a coin or knife. White quartzite is also a little more brown than marble.

Rock type: Non-foliated, contact and regional metamorphic
Texture: Even, medium-grained; sometimes granoblastic (grains are roughly shaped but even-sized)
Structure: Most small structures are obliterated by metamorphism, though bedding from the protolith may be preserved
Colour: White, grey, reddish
Composition: Tightly interlocking grains of quartz, with a little feldspar and mica
Protolith: Sandstones and quartz-rich conglomerates
Temperature: High
Pressure: Low to high
Notable occurrences: Islay, Grampians, Scotland; Anglesey, Wales; Norway; Sweden; Taunus, Harz, Germany; Wallis, Switzerland; Steiermark, Tyrol, Austria; North and South Carolina

Stone aggregates
Virtually every construction project in the world, from the simplest house to the biggest suspension bridge, relies on 'aggregate' – the small chunks of rock that are cemented together to make bricks, concrete, asphalt and various other building materials. The average house contains over 50 tonnes/49.2 tons of aggregate. Some aggregates are readymade from sand and gravel deposits. Most are crushed rock – and the choice of rock is crucial. Soft rocks such as shale are really usable only for cement. The main hard rocks are basalt, gabbro and granite, limestone, gritstone and sandstone and the tough metamorphic rocks hornfels, amphibolite and gneiss. Road aggregates must not just be tough; they must be resistant to polishing by tyres, since this makes them slippery when wet, and must allow bitumen to stick to them. This rules out quartz-rich rocks such as granite. As a result road chippings tend to be limestone, basalt, hornfels or amphibolite. A furnace quarry at a stone aggregate mine is shown above.

Granofels and charnockite

Granofels is one of the few non-foliated rocks to form under relatively high temperatures and pressures. This combination is generated only deep in the crust by tectonic forces that operate on a grand scale, so granofels is a product of regional, rather than contact, metamorphism. It is formed mostly from the granite family of rocks, or occasionally from thoroughly reconstituted clays and shales. Charnockite is a particularly widespread form of granofels. It was named by geologist T H Holland in 1900 after the tomb of Job Charnock, the founder of Calcutta, in St John's Church in Calcutta, India, which is made of this rock. Charnockite was once thought to be igneous, but it is now known to be metamorphic since despite the high temperatures and pressures, the original protolith never actually melted.

Rock type: Non-foliated, regional metamorphic
Texture: Coarse-grained
Structure: Most small structures are obliterated by metamorphism
Colour: Dark grey. Feldspar crystals may be dark green, brown or red; quartz may be bluish; hornblende may be brown to green.
Composition: Charnockite is made mostly of feldspar and quartz, but also contains the orthopyroxene hypersthene, plus hornblende and often pyrope garnet
Protolith: Granitoids and altered shales and clays
Temperature: High
Pressure: High
Notable occurrences: Scotland; Norway; Sweden; France; Madagascar; southern India; Sri Lanka; Brazil; Baffin Island; Labrador; Quebec; Adirondacks, New York

Identification: Granofels is a dark grey, coarse-grained rock mottled with brownish feldspar and greenish hornblende crystals.

MAFIC METAMORPHIC ROCKS

When mafic igneous rocks such as basalt are subject to regional metamorphism, increasing heat and pressure progressively changes them from greenstone to greenschist, and then to amphibolite and finally to eclogite. This sequence can often be traced in the landscape, at right angles to the direction of the original pressure when the rock was metamorphosed deep underground.

Greenstone and greenschist

Greenstones are often very ancient indeed, and bands of greenstone rock called greenstone belts are found wrapped around granite in cratons, the billions-of-years-old cores of continents. Neither greenstone, nor the related greenschist, are single kinds of rock. Instead, greenstone encompasses any metamorphosed mafic igneous rock turned greenish by the presence of chlorite, epidote or actinolite. Greenschist is similar but foliated, marked by schist-like stripes. Under mild pressure, basalt simply recrystallizes to greenstone, leaving structures such as pillows and cavities intact. Further compression breaks these structures down to create the foliation of greenschist. Greenschist is also the name of one of the facies of metamorphic rock, and includes greenstone. The greenschist facies is the assemblage of minerals that is formed by low-grade regional metamorphism – low temperatures (300–500°C/572–932°F) and only moderately high pressure. In greenschist facies, minerals such as albite, epidote, chlorite, actinolite, titanite and pumpellyite totally or partially replace the major minerals in the original igneous rock such as pyroxene and plagioclase.

Identification: Greenschist's green colour is its most distinctive feature. Like many metamorphic rocks it has slightly sparkly crystalline appearance. Unlike greenstone, greenschist is slightly foliated, with signs of the banding called schistosity.

Rock type: Non-foliated, low-grade regional metamorphic
Texture: Very fine-grained
Structure: Phenocrysts, cavities and pillow structures from the original volcanic rock may be preserved
Colour: Greenish
Composition: Mainly actinolite, with other epidote group minerals such as chlorite
Protolith: Mafic igneous rocks such as basalt (greenstone), or shale (greenschist)
Temperature: Low
Pressure: Moderate
Notable occurrences: Norway; Atlas, Morocco; Barberton, South Africa; Pilbara, Western Australia; Northwest Territories; Manitoba; Quebec; Ontario; Cascades; Rockies

Amphibolite

Identification: The high pressures and temperatures that form amphibolites mean their texture is distinctively metamorphic. Crystals have a unique contorted form called crystalloblastic, which can be created only by high-grade metamorphism.

This is a coarse-grained rock composed mostly of plagioclase and hornblende. Strictly speaking, it is non-foliated, but geologists may call any plagioclase-hornblende rock amphibolite, whether foliated or not. Amphibolite is also one of the metamorphic facies, encompassing the assemblage of minerals that form in any rock under the huge pressures and moderate temperatures typical deep down during mountain building. Extreme conditions like this turn amphibolites into some of the toughest of all rocks, which is why they are often used for building roads. Some amphibolites are metamorphosed from dykes and sills cutting clean across softer sedimentary rocks. The tremendous pressure and heat that alters these intrusions to amphibolite transforms even the softest surrounding sediments into schists and gneisses. However, amphibolite is so resistant to shear stress that it remains unfoliated and survives intact as fragments within the other metamorphosed rocks.

Rock type: Non-foliated, regional metamorphic
Texture: Medium- to coarse-grained. Sometimes contains porphyroblasts of garnet.
Structure: Hornblende crystals may be aligned, giving weak foliation
Colour: Black, dark green, green, streaked white, or red
Composition: Hornblende amphibole and plagioclase feldspar, plus mica almandine garnet and pyroxene
Protolith: Mafic, intermediate igneous rocks: basalt, andesite, gabbro, diorite
Temperature: Medium
Pressure: High
Notable occurrences: Donegal, Connemara, Ireland; Grampians, Scotland; Thuringia, Saxony, Germany; St Gothard Massif, Switzerland; Hohe Tauern, Austria; Quebec; Arizona; Adirondacks, New York

Eclogite

Eclogites are among the rarest of metamorphic rocks, but they are also among the most interesting. They are very striking-looking rocks, typically made of red pyrope or almandine garnets embedded in a green pyroxene called omphacite. No other rock is so often full of interesting crystals and minerals.

There is little doubt that they formed under extreme conditions. The assemblage of minerals in eclogite, called the eclogite facies, could have formed only under high temperatures and pressures. They are closely linked with basalts, and a few geologists have argued that they are not metamorphic at all, but formed directly from basalt magmas deep underground in the Earth's mantle. Eclogites are never very large. Although there are instances of isolated blocks measuring 100m/328ft across in metamorphic rocks, most are xenoliths – chunks of foreign stone swept up from the lower depths in magmas. Xenoliths such as this often occur in diamond-bearing kimberlite and lamproites, and the diamonds are usually embedded in the eclogites themselves. It is now thought that there are actually three different types of eclogite, each forming in a different way. There are those that occur as xenoliths in kimberlite and basalt, as in Hawaii's Oahu crater. These formed at extremely high pressures and temperatures at least 100km/ 62 miles down in the mantle. Secondly, there are those that occur in bands and lenses in the midst of the most extremely metamorphosed gneiss, like those in west Norway and the Dabie mountains of China. The third type of eclogite occurs as blocks or bands in subduction zones along with blueschist, as in the Greek islands of the Cyclades. These crustal eclogites formed at lower temperatures and pressures. One theory is that these formed from massive gabbros in conditions where there was little water present; others suggest they formed from deeply subducted basalt crust.

Identification: Tough, dense and coarse-grained, eclogite is basically green and can look almost like solid gelatin. Often it is studded with large red porphyroblasts of garnet like this specimen.

Rock type: Non-foliated or foliated, regional metamorphic. May also be igneous, forming from basaltic magma.
Texture: Medium- to coarse-grained. Often contains porphyroblasts of garnet or pyroxene.
Structure: Very high density, massive, occasionally foliated
Colour: Greenish, reddish, or green with red spots
Composition: Dominantly omphacite pyroxene and almandine-pyrope garnet, with no plagioclase. Also includes quartz, kyanite, orthopyroxene, rutile, pyrite, white mica, zoisite and occasionally coesite. Xenoliths may contain diamonds.
Protolith: Basalt or gabbro, marl
Temperature: Xenoliths in kimberlites, lamproites and orangeites above 900°C/1,652°F; eclogite lenses and xenoliths in ancient gneiss terranes 550–900°C/932–1,652°F; in blueschists near ocean trenches less than 550°C/932°F
Pressure: High
Notable occurrences: Glenelg, north-west Scotland; Greenland; West Norway; Saxony, Bavaria, Germany; western Alps, Switzerland; Carinthia, Austria; Apennines, Italy; Cyclades, Greece; DR Congo; South Africa; Botswana; Namibia; India; Borneo; Dabie Mts, central China; Western Australia; north-west Canada; Oahu, Hawaii; California

The ancient hearts of continents
Although some rocks forming the continents are quite young geologically, all of them have a very, very ancient core, or several cores. These ancient cores are called cratons, and continents have grown around them over the ages to become the land masses they are today. The rocks in cratons are the oldest on Earth, dating back at least 2.5 billion years. Gneisses metamorphosed from the volcanoes that created the first land masses are the oldest rocks. The Acasta gneiss of northern Canada is almost four billion years old. Almost as old are the distinctive greenstone belts, famous from Barberton in South Africa, from Pilbara in Australia and from northern Canada (above). Between 2.5 and 3.5 billion years old, these are rock islands of twisted greenstone wrapped around granite. Their origins are the subject of debate, but traces of pillow lavas are found in the greenstone, so many geologists think it may be pieces of ancient sea floor pushed up by a granite intrusion in ancient rift.

Retrograde eclogite: As eclogite xenoliths are brought nearer the surface, the minerals in them may sometimes be changed by retrograde metamorphism when they are affected by decreasing temperatures and pressures. Minerals such as amphibole may replace garnet and pyroxene.

MARBLE

Snowy white or cream with an extraordinary inner glow, marble is without doubt the most beautiful of all stones, cherished since the days of Ancient Egypt. For sculptors it is the finest of all stones, carved into shining statues such as Bernini's Ecstasy of St Theresa *and Michelangelo's* David. *Polished marble slabs have been used to face buildings from the Taj Mahal to the most modern skyscraper.*

Carrara marble

Identification: Pure marble is white but even pure marble can be stained grey by specks of graphite, or diopside like this. However, grey marble may be bleached snow white by contact with a hot intrusion.

Carrara marble: Those from the quarries near Carrara in Tuscany, Italy, are the most prized marbles of all. These snow-white rocks are almost pure calcite and were cherished not just by the great sculptor Michelangelo, but also in Roman times. The marbles occur in four main valleys in the Apennine mountains around Carrara.

For builders and sculptors, the word 'marble' covers a wide range of rocks. Limestones, serpentines and even quartzites may be called marble if they are pale in colour and can be carved or polished. For geologists, though, marble is a very specific kind of rock, made metamorphically under specific conditions. Even so, there is ambiguity. Some geologists describe as marble any rock that is metamorphosed from carbonate rocks. This includes metamorphosed dolomite, made from magnesium carbonate. A few geologists suggest that only rocks metamorphosed from pure calcium-rich limestones should be called marble.

This kind of true marble formed when beds of limestone were buried deep in the crust and altered by the heat of the Earth's interior and the pressure of overlying rocks. Marble is often brought up from deep mountain roots and continental collision zones interlayered with other ancient metamorphic rocks such as phyllites, quartzites and schists. Marble can be formed by contact as well as regional metamorphism, and small outcrops develop where hot granite intrusions have pushed their way into pure limestone beds.

Like metaquartzite, marble is made from reformed crystals of the same mineral as its protolith. During metamorphism, the calcite in limestone recrystallizes in larger, formless grains. Pore space between grains disappears, and grain and cement blur into one, leaving a tightly interlocking mass of calcite grains. The grains are odd shapes, and the texture can look sugary or even like the cracked pattern on ancient glazed porcelain. But this dense, uniform texture is just what makes marble so beautifully smooth for sculpture. The stone even seems to glow because it is slightly translucent and allows light to penetrate through the surface grains and reflect off internal grains.

Marble is soft enough to carve, but is tough enough to survive quite well in dry conditions. However, it is easily corroded by acid rain. Large masses of marble can be weathered into the same karst formations as limestone, and marble walls and statuary become pitted over time.

Rock type: Either foliated or non-foliated, low-grade regional metamorphic or contact metamorphic
Texture: Medium- or coarse-grained, clearly visible to the naked eye. Even-grained, often sugary in appearance. Translucent in slabs up to 30cm/12in thick.
Structure: Old bedding structures and even fossils are occasionally preserved. More often, though, marble is evenly massive. Pure marble is rarely foliated, but because it flows under high pressure, coloured minerals may be stretched out to give highly contorted stripes.
Colour: Occasionally pure white, but often stained different colours by minerals in the protolith. Pyroxene turns it green; garnet and vesuvianite turn it brown; and sphene, epidote and chondrodite turn it yellow. Other minerals and the process of metamorphism usually add waves, flecks, grains and stripes of colour.
Composition: Mainly calcite or, in dolomitic marble, dolomite. Additional minerals include quartz, muscovite and phlogopite mica, graphite, iron oxides, pyrite, diopside and plagioclases such as albite, labradorite and anorthite. Other minerals include scapolite, vesuvianite, forsterite, wollastonite tremolite, talc, chondrodite, brucite, apatite, sphene, grossular garnet, zoisite, tourmaline, epidote, periclase, spinel, pyrrhotite, sphalerite and chalcopyrite.
Protolith: Limestone. Carbonate limestone gives pure marble; dolomitic limestone gives dolomitic marble.
Temperature: Low to high
Pressure: Low to high

Marble varieties

Pure marble is white, and made mostly of calcite with minor traces of other minerals. The commonest additional minerals are small rounded grains of quartz, scales of pale muscovite and phlogopite mica, dark, shiny plates of graphite, iron oxides and pyrite. Different metamorphic conditions and different minerals in the original limestone can give it all kinds of different colours and patterns. Often these impurities are smudged out into wonderful whirling streaks, like ripple ice cream, as the rock flowed slightly during metamorphism. Common additional minerals are green diopside, pale green actinolite, plagioclase feldspars and many more (see data panel). Sometimes the entire mass of marble may be stained by impurities. Pyroxene turns marbles green. Garnet and vesuvianite turn it brown. Sphene, epidote and chondrodite turn it yellow. Graphite can turn marble grey or even black.

Once formed, marble can often be altered by both chemical and physical stresses. As it is attacked chemically, minerals such as hematite may develop, staining it red, while limonite stains it brown and talc stains it green. A particularly attractive alteration is when the marble is coloured by patches of green or yellow serpentine altered from diopside and forsterite in the original marble. This variety is called ophicalcite or verd antique.

The rock called onyx marble is not actually marble at all. Neither is it onyx, which is banded chalcedony. It is rings of calcite deposited from cold mineral-rich solutions around springs in crevices and caves, often as stalagmites. Onyx marble is also called alabaster and was widely used for carving in the ancient world. Reddish Siena marble from Tuscany is onyx marble, as is the Algerian marble used in the buildings of Carthage and Ancient Rome.

Dolomite marble: Dolomite marble is not true marble since it was metamorphosed from doleritic limestone and is made mostly from dolomite (magnesium carbonate) rather than calcite (calcium carbonate). This usually makes it a little greyer.

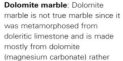

Ophicalcite: This marble gets its name, like 'serpentine', from the classical word for snake, and is basically serpentinized marble. Exposure to chemical attack changes forsterite and diopside to serpentine, giving the rock an appearance of snakeskin.

The artist's stone
Because of its softness, glow and beautiful colours, marble has been prized as a stone for sculpture since Egyptian times. Pentelic marble from Attica was the luscious white stone that Ancient Greek sculptors such as Phidias and Praxiteles used to make their wonderful statues, the first lifelike carvings of people. The famous Elgin marbles that once adorned the Parthenon in Athens were made of Greek marble, too. In the Middle Ages, Michelangelo carefully chose a block of pure white marble from the Carrara quarries in Tuscany, Italy, to make his great statue of David. Antonio Canova chose the same white stone for his famous *Three Graces*, and many others. Marble has been used even more widely to face buildings, ever since the Romans discovered how to stick it to walls with cement. So many of the buildings of Ancient Rome were covered in marble that the city shone even at night. Today, buildings such as Washington's National Gallery are clad in marble.

Notable occurrences:
Devon, England; Connemara, Ireland; France; Spain; Fichtelberg, Germany; Tyrol, Austria; Tyrol, Tuscany, Italy; Wallis, Switzerland; Talledega County, Alabama; Harford County, Maryland; Vermont; Georgia.

Pure marble: Mount Pentelicus (Attica), Greece; Carrara, Massa and Serravezza (Tuscany), Italy; Bergen, Norway; Alabama; Georgia; Maryland; Vermont; Yule, Colorado.

Dolomite marble: Glen Tilt, Scotland; Norway; Sweden; Fichtelberg, Germany; Steiermark, Austria; Tyrol, Italy; Karelia, Russia; Utah.

Ophicalcite: Sutherland, Scotland; Connemara, Ireland; Mona (Anglesey), Wales; Fichtelberg, Germany; Wallis, Switzerland; Alps, France; Piedmont, Italy; Estramadura, Portugal.

Onyx and stalagmite marble: Siena (Tuscany), Italy; Oued-Abdallah, Algeria; Tecali, Mexico; El Marmol, California.

Black marble (non-metamorphic limestone): Kilkenny, Galway, Ireland; Ashford (Derbyshire), Frosterley (Yorkshire), England; Shoreham, Vermont; Glen Falls, New York.

FOLIATED METAMORPHIC ROCKS

High or moderately high pressures during metamorphism can create rocks with distinct layers called foliation, including slate, schist and gneiss. Foliation makes some rocks stripy, like mylonite, migmatite and glaucophane schists, and makes others, like phyllite, liable to split into thin sheets. Foliation means either that some minerals have been separated into bands, or that crystals have been aligned in parallel.

Mylonite

Sometimes, the huge forces involved in crustal movement literally tear rocks apart and drag the broken edges past each other. Near the surface, rock is shattered into angular fragments along these fault zones and ultimately crushed to powder. Deep down, however, the heat of the Earth's crust makes rock too soft and plastic to break. So when rocks are sheared, they smear out like toffee to form streaky rocks called mylonites. Softer materials recrystallize as minute grains, while a few more robust larger crystals may survive, crushed and reduced within this fine matrix. Because the larger crystals have not recrystallized, they are called porphyroclasts, not porphyroblasts. It was once thought that the fine grains in mylonite were simply pulverized; however, it is now known that they are actually new crystals that form under the strain, in a process called syntectonic recrystallization. The word mylonite was coined by Charles Lapworth in 1885 to describe the streaked rock he found in the Moine Thrust fault zone of the Scottish Highlands. Now it is used to describe any rock with a smeared-out streaky texture like this, and bands can be anything from 1–2cm/ 0.4–0.8in to 2–3km/1.2–1.9 miles thick.

Identification: With their streaky, smeared-out texture, mylonites are quite easy to recognize, but there are many kinds of mylonite, including protomylonite in which many of the grains are porphyroclasts – pulverized but not recrystallized – and ultramylonite in which there are no porphyroclasts left at all.

Rock type: Foliated, dynamic metamorphic
Texture: Smeared-out, streaky texture with larger, but still tiny porphyroclasts in a very fine-grained matrix
Structure: Mylonites sometimes but not always split along the direction of the streaks
Colour: Varied
Composition: Varies with the original rock, but the matrix is typically quartz and carbonate, with feldspar and garnet porphyroclasts
Protolith: All kinds of rock
Temperature: Low
Pressure: High shear
Notable occurrences: Moine Thrust, NW Scotland; Alps, Switzerland; Turkey; Deccan Traps, India; Ross Sea, Antarctica; Canadian Shield; Sierra Nevada, California; Blue Ridge, Virginia; Adirondacks, New York

Migmatite

Migmatites are often the most extremely metamorphosed of all rocks, forming deep down in the continental crust under even greater pressure and heat than gneiss. Indeed, conditions are so hot that the rock partially melts. Minerals that melt at low temperatures liquidize, turning into igneous rock. Migmatites were first identified in 1907 by Finnish geologist J J Sederholm, who named them after the Greek word *migma* for mixture. The name is apt for migmatites are really a mixture of metamorphic and igneous rock. They usually consist of dark gneiss, schist or amphibolite striped by bands of leucocratic (pale-coloured) rock such as granite. At first migmatites were thought to be just pockets within gneiss. Now a few geologists think they may also be the last vestiges of rock that melted to form a granite magma. Some geologists argue that the pale bands have been intruded into the rock from an external source, rather than melting in situ. The gneiss portion is therefore older and called the 'palaeosome'.

Identification: With its distinctive humbug stripes of dark metamorphic rock and pale igneous rock, migmatite is usually fairly easy to identify.

Rock type: Foliated, regional metamorphic
Texture: Medium-grained
Structure: Alternating stripes of dark metamorphic and pale (leucocratic) igneous rock
Colour: Varied
Composition: Typically gneiss with granite. Can also be schist or amphibolite.
Protolith: May be all gneiss (or schist or amphibolite) or may be gneiss and granite
Temperature: Very high
Pressure: High
Notable occurrences: Sutherland, Scotland; Scandinavia; Auvergne, France; Black Forest, Bavaria, Germany; Cyclades, Greece; Lake Huron, Ontario; Adirondacks, New York; New Jersey; Washington

Phyllite

When mudstone and shale are subjected to mild metamorphism, their crystals line up perpendicular to the direction of pressure, and the rocks turn to slate. If the pressure and heat become a little more intense, they turn to phyllite. Even more heat and pressure turns phyllite to schist. Phyllite gets its name from the Latin for 'leaf-stone', and like slate it is characterized by laminations similar to the leaves of a book. Like slate, phyllite is made from very tiny grains of mica, chlorite, graphite and similar minerals, which grow flat at right angles to the pressure. Yet while slate looks dull, phyllite almost glitters because the extra heat and pressure creates thicker flakes of mica, especially the muscovite mica sericite. This silky sheen is called phyllitic lustre. In phyllite, the leaves are so compressed that it does not split into sheets nearly as well as slate, especially in its most highly metamorphosed, protophyllite, form. All the same, it is sometimes used, like slate, as roofing tiles.

Identification: Like slate, phyllite is distinctly layered. The layers, though, are not completely flat as in slate, but slightly wrinkled. These 'crenulations' make phyllite look like crepe. Phyllite also has a silky, silver lustre quite unlike slate's drab grey.

Rock type: Foliated, regional metamorphic
Texture: Fine-grained with porphyroblasts
Structure: Marked laminations at right angles to pressure. May have slaty cleavage and split into sheets as thin as 0.1mm/3.9mil. When metamorphism has gone further, the cleavage is only apparent and the rock won't split. May show minor folds and corrugations.
Colour: Silver grey, greenish
Composition: Mostly sericite mica and quartz
Protolith: Shale, mudstone
Temperature: Moderate
Pressure: Moderate
Notable occurrences: Donegal, Ireland; Grampians, Scotland; Anglesey, Wales; Cornwall, England; Scandinavia; Vosges, France; Fichtelberg, Bavaria, Harz Mts, Germany; Alps, Switzerland; Connecticut; New York; Appalachians

The Moine Thrust
The discovery of the Moine Thrust along the coast of Sutherland, north-west Scotland, in 1907 was a key moment in the history of geology. Geologists were already familiar with simple thrust faults. These are shallow reverse faults created when the crust is squeezed, forcing one block of rock up over another. But the Moine Thrust is not a single thrust. In fact, it is a complex belt of thrusts. We now know that such belts develop when tectonic plate movement repeatedly forces layers of crust up and over each other, then pulls them back, creating a complex, broken, multi-layered formation. At Moine, this all happened between 410 and 430 million years ago, when Scotland was crushed by opposing tectonic plates, creating a thrust belt stretching 180km/112 miles from the Moine Peninsula to the Isle of Skye. Similar thrust belts have now been discovered along the edges of fold mountain ranges all around the world. They are often characterized by complex bands of mylonite rock.

Glaucophane schist: blueschist

Glaucophane schist is a rock turned blue by the amphibole mineral glaucophane. It is also called blueschist, but it is one of a variety of rocks that form in similar conditions known as the blueschist facies. Not all blueschist facies rocks are blue, but laboratory experiments have shown they all form when pressures are very high but temperatures are low. This is a surprising combination, since high pressures usually go hand in hand with high temperatures, forming greenschists. Geologists now believe the answer is that blueschists form in subduction zones. The theory is that as a cold basaltic ocean slab is shoved deep into the mantle, a wedge of material, called an accretionary wedge, is scraped off and pushed back to the surface by slab material descending behind. This all happens so quickly in geological terms that the rock in the descending slab is squeezed hard, altered to blueschist then lifted back on the surface before it has time to heat up.

Identification: Schistose banding gives away glaucophane as a schistlike rock. A bluish tinge may establish its identity.

Rock type: Foliated, regional metamorphic
Texture: Fine, medium-grain
Structure: Weakly schistose. May show folding.
Colour: Bluish, light violet
Composition: Mostly glaucophane or lawsonite amphibole or epidote, quartz, or jadeite, with garnet, albite, talc, zoisite, jadeite and chlorite
Protolith: Usually basalt or dolerite, but may be mudrock
Temperature: Low
Pressure: High
Notable occurrences: Anglesey, Wales; Channel Islands, England; Spitsbergen, Norway; Calabria, Tuscany, Val d'Aosta, Italy; Alps, Switzerland; California

Schist

Mica

SLATE

The word slate is sometimes used to describe any stone that splits into flat slabs and is used as roofing tiles. True slate, however, is a very distinctive dark grey, brittle metamorphic rock that flakes into smooth, flat sheets. It is created when shales and clays are altered by low-grade regional metamorphism at low temperatures and moderate pressures.

Slate

Getting its name from the Old German word for break, slate is essentially metamorphosed mudrock that has been strongly compressed deep underground, but in a low-grade regional metamorphic environment, away from the most intense metamorphism. Conditions such as these are found deep at the root of fold mountains, where the convergence of tectonic plates slowly but surely crushed rocks deep down. Most slates occur in old mountain chains, like the Appalachian Mountains of the USA, or Snowdonia in Wales. They tend to be Precambrian or Silurian in age. Occasionally, though, they form in more recent fold mountain chains, like the Alps.

Slate varies enormously in colour, though it is normally dark grey, or purplish or greenish grey. But it is always easy to recognize because of the way it cleaves into the flat sheets that make it so useful for roofing. This distinctive 'slaty' cleavage develops when mudrock is metamorphosed. As the rock is compressed, water is squeezed out and the rock is compacted. All the tiny clay and silt grains are not simply recrystallized as mica and chlorite but are reoriented at right angles to the pressure. This alignment occurs partly because just as the layered nature of clay crystals means they can be moulded, they can also be pressed flat, and partly because new crystals grow in this direction.

Metamorphism usually destroys most of the original sedimentary structures, so the cleavage planes are entirely unrelated to bedding planes and are probably at an entirely different angle. Fossils are usually destroyed by metamorphism, too. Only when the pressure is pretty much at right angles to the bedding are fossils preserved – though often rather dramatically flattened. Because slate's cleavage is produced by the same forces that fold mountains, slate cleavage usually clearly marks out the pattern of folding in the formation and the direction of compression. This makes it a very useful rock when studying the tectonic history and structural geology of an area.

Identification: A dull dark, smooth grey, turning shiny black when wet, slate is a distinctive rock, even before it is broken. But its tendency to break into thin, flat sheets marks it out even more clearly. No other rock has such a distinctive cleavage.

Clayslate: This is a very mildly metamorphosed rock halfway between shale and phyllite. Some geologists regard it as a sedimentary rock, but it has been metamorphosed enough to prevent it absorbing water and swelling like shale. Clayslates never smell earthy when damp like shale, and rarely have any fossils. They also split like true slates.

Rock type: Foliated, low-grade regional metamorphic
Texture: Very fine-grained. The grains are so small that it is impossible to identify individual minerals even under a magnifying glass.
Structure: Slates are characterized by single, perfect flat, slaty cleavage that allows it to be split easily into sheets. Traces of bedding planes and other original protolith structures are sometimes revealed like a picture on the flat cleavage surfaces. Sometimes fossils are preserved but they are usually squeezed flat.
Colour: Grey, black, shades of blue, green, brown and buff. Limonite and hematite colour it brown; chlorite colours it green.
Composition: Mainly mica and chlorite, with quartz, pyrite and rutile. Minor minerals include calcite, garnet, epidote, tourmaline, graphite and dark carbonate minerals.
Protolith: Mudrocks (mostly shale) and volcanic tuff
Temperature: Low
Pressure: Moderate
Notable occurrences: Wicklow Mts, Ireland; Highland Boundary Fault, Scotland; Snowdonia, North Wales; Cornwall, England; Ardennes, France; Fichtelberg, Thuringia, Germany; South Australia; Brazil ('rusty' slate); New Brunswick; Nova Scotia; Ontario; Martinsburg, Pennsylvania; western Vermont; eastern New York; central Virginia; central Maine; northern Maryland

Spotted and chiastolite slate

Slate is the first stage in the sequence of rocks that develop when mudrock is progressively metamorphosed on a regional scale. Moderate heat and pressure on mudrock alters it to slate, but if the pressure increases it develops into phyllite. Even more intense pressure turns phyllite to schist. Just like sedimentary and igneous rocks, slate can also be cooked and altered by contact with a hot granite intrusion. Right next to the intrusion, slates lose their distinctive cleavage, and develop into splintery, tough hornfels. Further away, they retain their cleavage but develop dark or light round spots of altered minerals, typically white mica or chlorite. Spotted slates such as these usually contain minerals such as andalusite, garnet or, more rarely, cordierite. Andalusite in the form of chiastolite is especially characteristic of these spotted slates. Chiastolite slates contain distinctive large porphyroblasts of andalusite with dark crosses embedded in a light crystal. These crystals can often be 7–8cm/2.3–3.1in long. Commonly, in exposed slates, though, they are weathered to white mica or kaolin.

Spotted slate: Just as shale is metamorphosed to slate by pressure on a regional scale, so slate is metamorphosed locally to form spotted slate by contact with a hot intrusion.

Chiastolite slate: Chiastolite slate is a very distinctive rock. Like ordinary slate, though, it is tough yet brittle and breaks easily into flat sheets. Unlike ordinary slate, it is marked with porphyroblasts, created by the heat of contact with an intrusion. The porphyroblasts are pale crystals of andalusite, aligned randomly throughout the rock.

Rock type: Foliated, low-grade regional metamorphic
Texture: Very fine-grained, but studded with porphyroblasts of minerals such as andalusite and mica
Structure: Like ordinary slates, they are characterized by single, perfect, flat, slaty cleavage that allows them to be split easily into sheets
Colour: Grey, black, shades of blue, green, brown and buff. Limonite and hematite colour it brown; chlorite colours it green.
Composition: Mainly mica and chlorite, with quartz, pyrite and rutile. Minor minerals include calcite, garnet, epidote, tourmaline, graphite and dark carbonate minerals.
Protolith: Mudrocks (mostly shale) and volcanic tuff
Temperature: Moderate
Pressure: Moderate
Notable occurrences: Snowdonia, Wales; Skiddaw, Devon, England; Betic Cordilleras, Spain; Halifax, Nova Scotia; California

The slate industry

Slate may be very brittle and split easily into sheets, but it is actually a tough rock that is very resistant to the weather. This is why it is often seen in craggy outcrops in mountain regions, standing out darkly, almost black when it rains. The combination of slate's weather resistance, and the ease with which it can be broken into flat sheets makes it a superb light and durable roofing material.

Buildings have been roofed with slate for thousands of years. The famous slates of

North Wales have been used since at least Roman times. The Roman fort of Segontium (modern Caernarfon) had its tiles replaced with slate in the 4th century, while at nearby Caer Lugwy a fort had a slate roof two centuries earlier.

By the Middle Ages, Welsh slate was being shipped all around the British Isles, and maybe even abroad. When Chester castle was renovated in 1358 under the supervision of Edward the Black Prince, 21,000 slates were shipped from Wales to cover its roof. Many other castles and large houses used slate from the Welsh quarries at Cilgwyn and Penrhyn. Further north in Scotland, a thick, more durable kind of slate was used on the roofs of medieval houses in Edinburgh.

The use of slate remained fairly small-scale, though, until the 19th century and the Industrial Revolution. As cities in both North America and Britain grew rapidly, houses were built in millions, and each one of them had a slate roof. In Britain, the output of slates from the Welsh quarries expanded enormously. By 1832, they were digging out 100,000 tonnes/98,420 tons a year. Half a century later, almost half a million tonnes of slate were coming out of the Welsh

quarries alone. In North America, the situation was similar, with huge quantities of slate dug out of the quarries of Vermont and Pennsylvania to cover the roofs of millions of houses.

Demand for slate has declined dramatically since its 19th-century peak, as natural slate has been replaced by cheaper, more regular, mass-produced artificial tiles. The Welsh quarries now produce less than 5 per cent of what they did at their height, and Vermont quarries even less.

Slates are still made today by hand by slaters in the same way they have been for centuries. Large blocks are dug out, then cut with a saw across the grain into sections slightly longer than the finished tile. Then the blocks are 'sculped' or split into slabs with a mallet and a broad-faced chisel. Finally, the slabs are split into tiles using a mallet and two chisels, and trimmed down to exactly the right size and shape. A last touch may be to punch two nail holes in one end. These holes allow the slate to be fixed quickly in place by the roofer. This last task requires some care if the tile is not to split, and relies on the slater's estimate of the particular qualities of the slate.

SCHISTS

The resilient rock that provides a firm base for Manhattan's towering skyscrapers, schist is the most extreme form of the regional metamorphism of mudrocks. When pressures on slate and phyllite climb, and temperatures rise above 400°C/752°F, the rocks completely recrystallize and reform as schist. All schists form this way, but there are many kinds, depending on the minerals they contain.

Schists

Folded schist: Sometimes the layers in schist can be intensely contorted even on a small scale.

Schist is one of the most striking of all rocks, with its distinctive layering or 'schistosity'. Schists mostly develop deep in the roots of mountain ranges as they are being folded. Here pressure and temperatures reach the point where the original clay minerals in mudrocks, slates and phyllites are completely broken down. The chemicals in them then reform as larger crystals of minerals such as mica and chlorite. Schists are therefore usually medium- to coarse-grained rocks, with crystals that, though not as well defined, can be as large as those in granite.

Schistosity develops only partly because the different minerals separate into layers. The main reason is that the continuous squeezing and shearing of the rock during metamorphism allows crystals only to grow in one plane, at right angles to the pressure. Because the crystals that develop are tabular flakes such as mica, or needle-like crystals such as amphiboles, this layering effect is even more marked. Whenever schist is split, it tends to break along layers of mica within the rock, giving the slightly misleading impression that it is entirely made of glistening mica. Quartz is also an important constituent, but the quartz layers tend only to be seen clearly when the rock is cut across the grain.

Although schist breaks into sheets like slate, and is sometimes used as roofing tiles where slate is scarce, it does not cleave quite as easily. The extreme heat and pressure mean the layers are more tightly knitted together. They are often also slightly wrinkled, or 'crenulated', similarly to phyllite but even more so. In some schists, these contortions can become quite dramatic.

Although they develop mainly from mudrocks, they can form from any rock that contains the right constituent minerals to make mica. Like gneiss, they are found primarily in areas of ancient rock. Most date back to at least Precambrian times (older than 570 million years). Schists do also occur in more recent formations, though, such as in the young fold mountain belts of the Alps and Himalayas.

Garnet schist: Under fairly intense metamorphism, large porphyroblasts of garnet may develop in mica schist. Garnets 10mm/ 0.4in or more across can sometimes be seen, as in this garnet schist, and may be big enough to chip out and use as gems. The schistose layers tend to bend around the garnet like a stream flows around rocks. The garnets are typically iron-rich, pink almandine metamorphosed from pyroxenes and other minerals.

Chlorite schist: In chlorite schist, weakly developed schistosity is created by flaky green chlorite crystals and fine green needles of actinolite which often form radiating clusters. This rock forms under relatively mild, 'greenschist' regional metamorphic conditions.

Chlorite schist
Rock type: Foliated, regional metamorphic
Texture: Fine to medium-grained, sometimes with porphyroblasts of albite or chloritoid. Not as markedly schistose as mica schist. Unlike gneiss, schist's crystal fabric is dominated by long 'planar' crystals. This is a useful clue to identity.
Structure: Typically folded, on a small or large scale
Colour: Greenish grey
Composition: Chlorite, actinolite, epidote, talc, glaucophane and albite feldspar. Little or no mica and quartz.
Protolith: Mainly mudrocks such as shale
Temperature: Moderate
Pressure: Moderate
Notable occurrences: Argyll, Scotland; Lake Tauern, Austria; Tyrol, Piedmont, Lombardy, Italy; Sierra Nevada, California

Garnet schist
Rock type: Foliated, regional metamorphic
Texture: Medium- to coarse-grained. Garnet porphyro-blasts common. Well-developed schistosity.
Structure: Typically folded, on a small or large scale
Colour: Black, brown, reddish
Composition: Garnet, plus biotite and muscovite mica and quartz. Garnet schist also contains the same range of other minerals as mica schists.
Protolith: Mainly mudrocks such as shale
Temperature: Moderate to high
Pressure: Moderate to high
Notable occurrences: Connemara, Ireland; Scotland; Scandinavia; Alps, Switzerland; Black Forest, Germany; Quebec; Duchess Co, NY; New Hampshire

Mica schists

The most common schists are mica schists. Flakes of mica give them the most marked schistosity, and a real shine. The flakes are typically about 0.5mm/0.02in thick and can often be prised off with a knife. There is plenty of quartz in mica schist too, often concentrated in mica-poor layers, and a fair amount of albite feldspar. Sometimes red garnet or green chlorite crystals are visible. The mica in mica schists can be muscovite, sericite or biotite. Biotite is usually brown; muscovite and sericite are pale coloured and called white mica. If a white mica is fine-grained, it is called sericite; if it is coarser it is called muscovite. Most mica schists contain all three, but one usually predominates. Muscovite and sericite schists develop where metamorphism is of moderate intensity, and are often associated with greenschists and phyllites. Biotite develops partly at the expense of muscovite (and chlorite) when metamorphism becomes more intense. Even more intense metamorphism edges the schist towards garnet schists.

Biotite schist: Biotite schist is brown and dark, but still has the mica gleam.

Muscovite schist: Muscovite schists, with sericite schists, are the lightest-coloured schists, coloured by white mica. Unlike sericite schist, the grains of mica in muscovite schist are clearly visible to the naked eye.

Mica (muscovite, sericite or biotite) schist
Rock type: Foliated, regional metamorphic
Texture: Medium to coarse-grained, sometimes with porphyroblasts. Thin schistose layering of mica flakes always marked.
Structure: Typically folded, on a small or large scale
Colour: Light grey, greenish (biotite schist is browner)
Composition: Quartz and mica (usually muscovite), plus kyanite, sillimanite, chlorite, graphite, garnet, staurolite
Protolith: Mainly mudrocks such as shale
Temperature: Moderate
Pressure: Moderate
Notable occurrences: Connemara, Ireland; Scotland; Scandinavia; Alps, Switzerland; Black Forest, Germany; Quebec; Duchess Co, NY; New Hampshire

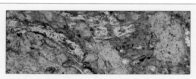

Metamorphic facies
It is impossible to tell from a single mineral at what pressure and temperature a metamorphic rock formed, but you can tell from the groupings of minerals known as 'facies' – assemblages that form in certain conditions. These facies include: zeolite, greenschist, amphibolite, blueschist, eclogite, granulite and various hornfelses. Although most facies are named after a rock containing one of these groups of minerals, different rocks can be in the same facies. Amphibolite and hornblende schist both form in amphibolite facies, for instance, and the above thin section shows a greenschist to amphibolite facies in meta-basalt, in which most of the green crystals are in fact hornfels amphibole. Although the link isn't always definite, hornfels, granulite and eclogite are high-grade; greenschist and amphibolite medium-grade; and zeolite and blueschist low-grade. Hornfels facies are linked to high-temperature contact zones; granulite to the high temperatures and moderate pressures typical of mountain roots; amphibolite to moderate temperatures and pressures in mountain roots and continental interiors; greenschists and zeolite to mild conditions under continental interiors; and blueschist to low temperatures and high pressures in accretionary wedges along subduction zones.

Amphibole schist

Amphibole schist is schistose like mica schist, but the layers are formed not by flakes of mica but by parallel bands of long, thin amphibole crystals. If the crystals are not parallel, the rock is not schistose and is called amphibolite. The amphibole in amphibole schist is usually hornblende, and the rock is called hornblende schist, but it can also be actinolite or tremolite. All amphibole schists are much richer in feldspar than amphibolite. Other minerals in hornblende schist include chlorite, epidote, pyroxene and garnet. Garnet often forms dark red porphyroblasts.

Dark amphibolite crystal

Amphibole (hornblende) schist
Rock type: Foliated, regional metamorphic
Texture: Medium- to coarse-grained, sometimes with porphyroblasts. Thin schistose layering of amphibole needles always marked.
Structure: Typically folded, on a small or large scale
Colour: Dark black or brown, often streaked or flecked with white or red
Composition: Hornblende (or actinolite or tremolite), plagioclase feldspar, quartz and biotite mica, plus pyroxene, epidote, muscovite mica and garnet
Protolith: Basalt and dolerite, as well as mudrocks
Temperature: Moderate
Pressure: Moderate
Notable occurrences: Connemara, Ireland; Scotland; Tyrol, Austria/Italy; St Gotthard Massif, Switzerland; Quebec; Mitchell Co, North Carolina

GNEISS AND GRANULITE

The word gneiss (pronounced 'nice') comes from an old slavonic word for 'sparkling', and that's exactly what gneiss does. Under a microscope it can be seen to be made of tightly packed, iridescent crystals forged by the most intense metamorphism of all. Gneiss and granulite are incredibly tough rocks and are found together in terranes that are the most ancient rock formations on Earth.

Gneiss

Identification: Gneiss can often be identified by its humbug stripes of dark and light minerals and its crystal fabric (see Texture, right). It is also incredibly tough, found in the most ancient and time-worn landscapes, such as the vast Canadian Shield and Scotland's Hebridean islands.

Almandine garnet

Granular gneiss: Granular gneiss is less distinctly banded than other gneisses. Its composition is often close to granite, and is made from high proportions of quartz, white and pink feldspar and white and dark mica.

Unlike schist, gneiss is not dominated by long, planar crystals. But it can often be the most markedly striped of all rocks, composed of alternating bands of light and dark minerals. In most cases these bands are just 2mm/0.08in thick, but they can be as wide as 1m/39in. The bands in gneiss are not like the layers in schist, which are made from sheets of mica, nor does gneiss split easily in the same way. The high temperatures it takes to create gneiss tend to destroy mica, and the banding is formed in an entirely different way – as minerals separate out and form into distinct bands. The light-coloured bands are formed by light-coloured, typically felsic, minerals such as quartz, feldspar and white mica (usually muscovite). The dark bands are made of dark, typically mafic, minerals like amphibole, pyroxene and biotite mica. Compositional bands like these form only at very high temperatures when minerals almost melt and so can move freely before recrystallizing. Gneisses form deep in the Earth in subduction zones or under the roots of fold mountains, and are bought to the surface only by massive tectonic movements, or the slow erosion of the overlying mountains.

Compositional bands in gneiss can also form when variations in the original rock survive through all the stages of metamorphism. A rock originally made of alternate narrow beds of shale and sandstone may be transformed by metamorphism into gneiss made of alternate bands of quartzite and mica.

Some gneisses, however, derive their bands in an entirely different way. These gneisses get their bands when thin floods of granitoid magma ooze their way in between layers of the protolith, or when there is local melting. These gneisses blend into migmatites, and have led some geologists to conclude that many ancient gneisses formed from granodiorite and tonalite magmas in this way. Such gneisses are closely linked to ancient greenstone belts, and often merge into them.

Gneiss is an incredibly tough rock, perhaps the toughest of all, and vast quantities of it have survived since the very early part of the Earth's history. Large areas of Greenland are made from gneisses at least three billion years old. The world's oldest known rock is Acasta gneiss from northern Canada, which has been dated to 3,900 million years ago.

Rock type: Foliated, regional or dynamothermal metamorphic
Texture: Medium- to coarse-grained. Marked by striking alternate light and dark bands. The light bands are often coarsely granular. The dark bands are finer-grained and, when they contain biotite mica, may be foliated like schist. Unlike schist and granulite, gneiss has a mix of long 'planar' crystals and 'equant' crystals, crystals that measure much the same in all directions. Schist has more planar crystals and granulite more equant crystals. This difference in crystal fabric is a useful clue to identity.
Structure: May be marked by large-scale as well as small-scale dark and light bands. Often folded. Very often criss-crossed with granite and pegmatite veins.
Colour: Greyish, pinkish, reddish, brownish, greenish with dark stripes
Composition: Varies with protolith, but typically abundant in feldspar and quartz and white mica, which forms the light layers, and biotite and hornblende which forms the dark layers. Other constituents include cordierite, garnet and sillimanite.
Protolith: Almost any other rock. Gneiss formed from igneous rocks is called orthogneiss; gneiss made from sedimentary rock is called paragneiss.
Temperature: High
Pressure: High
Notable occurrences: Lewis, Orkneys, Scotland; Greenland; Scandinavia; Vosges, Massif Central, Brittany, France; Bavaria, Erzebirge, Germany; Alps, Switzerland; Southern India; Thailand; Canadian Shield; Appalachians; Idaho

Granulite

Granulite is a tough, coarse-grained rock that, like gneiss, forms at very high temperatures and pressures. It is also the facies of metamorphic minerals that form under these extreme conditions, which tend to destroy mica and replace it with minerals such as pyroxene. The high pressure drives out any water, so the minerals that form are said to be anhydrous. It is thought that granulite formed at the base of the continental crust. Indeed, most of the underside of the continental crust is probably made of granulite. Granulite has mostly reached the surface either as small xenolith chunks in magmas, or when mountain ranges are worn away so far that their very roots are exposed. Most granulites are very ancient, and are found with gneisses in the granulite-gneiss terranes that contain the oldest rocks on Earth, dating back to billions of years ago.

Mineral identifiers
About a century ago, geologist George Barrow was investigating the rocks of the Scottish Highlands around Aberdeen (pictured above). Here shale, sandstone, limestone and mafic lava were crushed and folded during a powerful phase of mountain building that created the ancient Caledonian mountains. As he studied these ancient metamorphosed rocks, Barrow noticed how particular minerals appeared in rocks in a sequence across the landscape, reflecting different intensities of the metamorphosis of pelites (metamorphosed mudrocks). These 'index' minerals were, in order: chlorite, biotite, garnet, kyanite and sillimanite. Zones in which these index minerals appear are now called Barrovian metamorphic zones, and the boundary of each is marked by a line on a map called an isograd. Similar zoning was found running through andalusite, cordierite, staurolite and sillimanite. Zones in which these minerals are found are called Buchan zones. The Buchan sequence is thought to be created by lower pressures than Barrovian zones.

Rock type: Mostly non-foliated, regional or dynamothermal metamorphic
Texture: Coarse-grained. Often banded like gneiss but granulite has mostly 'equant' crystals (see Gneiss; Texture).
Structure: Marked by dark and light bands
Colour: Light, almost white
Composition: Pyroxene (diopside or hypersthene), quartz and feldspar, plus garnet, biotite, cordierite and sillimanite
Protolith: All kinds of rock
Temperature: High (>700°C)
Pressure: High
Notable occurrences: NW Scotland; Greenland; Finland; Aldan shield (Yakut), Siberia; Ukraine; Limpopo, South Africa; Hopeh, Liaoning, China; Yilgarn, Australia; Enderby Land, Antarctica; Canadian Shield; British Columbia; Adirondacks, NY; Beartooth Mts, Montana

Identification: Granulite is a hard, sparkling rock made of coarse, rounded interlocking grains of mostly pale minerals.

Augen gneiss and other gneisses

Gneisses can be distinguished by their protolith, such as granite gneiss and syenite gneiss, or by their characteristic mineral, such as biotite gneiss and garnet gneiss. Some gneisses may be distinguished by their texture, such as platy gneiss and augen gneiss. The word augen comes from the German for 'eye', and refers to large, oval or eye-shaped crystals in the rock. The crystals are typically alkali feldspar, in a matrix of quartz, feldspar and mica, but can be quartz, or garnet (in which case the rock is known as garnet augen gneiss). Each eye, or auge, can be up to 10cm/3.9in across. Feldspar augen typically contain inclusions of minerals such as biotite. The augen in a rock tend to be much the same size and shape, and it is thought that they are survivors from an earlier stage, with a core too big to be affected by the recrystallization that aligned other minerals in bands. The bands flow around them, like a stream around a rock. Sometimes, garnet augen rotate as they grow during metamorphism, creating 'snowball' garnets, containing spiral inclusions of other minerals.

Identification: Both schists and gneisses can contain augen, which are large, oval crystals of feldspar, quartz or garnet. Augen gneiss is more common, and the augen are usually surrounded by dark, banded gneiss layers, rather than silvery, flaky schist.

Rock type: Foliated, regional or dynamothermal metamorphic
Texture: Medium- to coarse-grained. Marked by large pale crystals or augen.
Structure: May be marked by large-scale as well as small-scale dark and light bands
Colour: Greyish, pinkish, reddish, brownish, greenish with dark stripes
Composition: Augen made of alkali feldspar or garnet. Matrix of feldspar and quartz forming light layers, and biotite forming the dark layers.
Temperature: High
Pressure: High
Notable occurrences: Lewis, Orkneys, Scotland; Greenland; Scandinavia; Vosges, Massif Central, Brittany, France; Bavaria, Erzebirge, Germany; Alps, Switzerland; Canadian Shield; Appalachians; Idaho

ROCKS ALTERED BY FLUIDS AND OTHER MEANS

Not all metamorphic rocks are formed directly by the heat of contact with an intrusion, or by heat and pressure deep within the crust. Skarns and serpentinites are formed by the interaction of fluids with the country rock – skarns by fluid heated by an intrusion, serpentinites by cold water. Halleflintas are altered volcanic tuffs and fulgurites are rocks formed by the intense heat of lightning strikes.

Skarn

Skarns are treasure troves of unusual and valuable minerals such as grossular garnet and ores of iron, copper, lead, tungsten and zinc. They are typically patches of metamorphosed rock around a granite intrusion, but the term 'skarn' covers a wide range of different rocks and mineral deposits that originated in a variety of different ways. Some geologists use the word skarn to describe any calcium- and silicate-rich metamorphic rock containing unusual minerals. Most prefer to describe skarns as only metamorphic rocks that form when limestones and dolomites are altered by contact with hot granite intrusions. Granite intrusions generate hot fluids carrying copious amounts of silicon, iron, aluminium and magnesium either emanating directly from the intrusion or cooked up as the intrusion heats groundwater in the limestone. Infiltrating the limestone, this rich brew alters minerals to calcium, iron and magnesium silicates. This process is really metasomatism, not metamorphism, because the minerals in the rock are replaced by others as they come into contact with the hot fluids.

Apatite

Orange calcite

Identification: Skarns are very piebald in appearance. They are characterized by large patches of minerals of different colours, formed as they were concentrated by the hot fluids that oozed through the limestone during metasomatism.

Rock type: Non-foliated, hydrothermal metasomatic
Texture: Fine-, medium- or coarse-grained
Structure: Occurs in small patches, with minerals concentrated in nodules, lenses and radiating masses
  Colour: Brown, black or grey but very variable
  Composition: Pyroxene, garnet, idocrase, wollastonite, actinolite, magnetite, epidote. Skarns host copper, lead, zinc, iron, gold, tungsten, molybdenum and tin ores.
Protolith: Mostly limestones and dolomites
Notable occurrences: Dartmoor (Devon), England; Central Sweden; Elba, Italy; Trepca, Serbia; Banat, Romania; Arkansas; Crestmore, California

Halleflinta

Identification: No metamorphic rock looks more like flint than halleflinta. It is very fine-grained, almost cryptocrystalline, and splinters like flint when hit with a hammer.

Halleflinta gets its name from the Swedish for 'rock-flint'. It is a very hard, flinty, metamorphic rock so fine-grained that it is hard to identify individual minerals even under a microscope. It is basically a very intimate mix of quartz, feldspar and other silicate minerals. It forms in similar conditions to gneiss and schist, under intense heat and pressure, and often occurs in association with them in Scandinavia, where it was first identified. But it contains none of the banding of gneiss, nor the schistosity of mica. Indeed, it is almost glassy in texture, and breaks into sharp splinters like flint. The reason for this difference is almost certainly due to its protolith. It is probably metamorphosed volcanic tuff, and halleflinta often retains signs of the original layers of volcanic debris as it settled after successive eruptions. Extra silica has usually got into the rock during metamorphism.

Rock type: Foliated, regional or dynamothermal metamorphic
Texture: Very fine- and even-grained, almost glassy, so that the rock splinters. May contain larger porphyroblasts of quartz.
Structure: Layering related to original volcanic deposit, but no schistosity or banding
Colour: Grey, buff, pink, green or brown
Composition: Quartz, feldspar, mica, iron oxides, apatite, zircon, epidote, hornblende
Protolith: Volcanic tuff
Temperature: High
Pressure: High
Notable occurrences: Sweden; Finland; Tyrol, Austria; Bohemia, Czech Republic; Galicia, Poland; Ukraine

Serpentinite

Serpentinization is a process that alters rocks, but it is not like other forms of metamorphism. It gets its name because it creates a rock flecked like snakeskin called serpentinite. This consists mostly of fibrous serpentine minerals such as chrysotile, antigorite and lizardite. In serpentinization, it is not heat and pressure that alter the minerals but heat and water. It affects mostly ultramafic rocks such as peridotite and dunite, although serpentinite can form from gabbro and dolomitic limestone, as well. What happens is that water infiltrates the rock and alters iron-rich minerals such as olivines and pyroxene to create serpentine minerals. The water is cool but the chemical reaction is exothermic and generates its own heat. It was once thought serpentinite was quite rare, and occurred only above subduction zones, or within small ultramafic intrusions. Now it is realized that serpentinites pretty much underlay the entire ocean floor, forming part of the ophiolite sequence, as olivine-rich magmas oozing up through the mid-ocean rift are serpentinized by sea water. Hydration (the uptake of water) makes serpentinites light so they well up in many places, creating undersea mountains. They also well up in the accretionary wedges above subduction zones.

Identification: Serpentinites are dark green to black and look very much like the snakeskin that earned them their name. They are often quite coarse-grained, and green serpentine crystals are usually easy to see.

Rock type: Foliated, hydrothermal metamorphic
Texture: Medium- to coarse-grained. Compact, dull, waxy. Fractures in splinters.
Structure: Often banded. Usually criss-crossed with veins of chrysotile serpentine.
Colour: Grey-green to black
Composition: Serpentine (chrysotile, lizardite, antigorite), olivine, pyroxene, hornblende, mica, garnet, iron oxides
Protolith: Peridotite, dunite, pyroxenite, and occasionally gabbro and dolomite
Notable occurrences: Lizard (Cornwall), England; Shetland Islands, Scotland; Pyrenees, Vosges, France; Liguria, Italy; Montana; Oregon; California; Maine; The Lost City, Mid-Atlantic seabed; Izu-Bonin-Marian seamounts, Pacific

The Lost City
Geologists have long known about 'black smokers', or hydrothermal vents (above). These are remarkable chimneys on the sea floor in the mid-ocean ridge. They are built up from deposits left by smoky clouds of sulphide-rich water that bubble up through the sea floor, superheated by magma. In 2001, oceanographers discovered an entirely different kind of smoker towering up from the Atlantic ocean floor. Forming what was dubbed the Lost City, these white smokers are made of carbonate, and develop in an entirely different way to black smokers. For a start, they form well away from the central ocean rift. More significantly, they are heated not by magma but by the heat generated from serpentinization reactions in peridotite rocks under the ocean floor surface. As sea-water infiltrates the peridotites, it not only turns them to serpentinite but also generates warm, alkali-rich waters that bubble up in white smokers to form brucite and calcite towers.

Fulgurite

Fulgurites are the most unusual and rarest of all metamorphic rocks. They get their name from the Latin for 'thunderbolt', and they are natural tubes or crusts of glass that form when lightning strikes. To fuse sand instantly into glass needs temperatures of 1,800°C/3,272°F, and lightning regularly reaches a searing 2,500°C/4,532°F. There are two kinds of fulgurite: sand and rock. Sand fulgurites form in the loose sand on beaches and in deserts. They are branching tubes that look like roots. They average 2.5cm/1in in diameter and can be up to 1m/39in long. Rock fulgurites are crusts or coats of glass that form when lightning strikes solid rock. Typically, they form branching marks across the rock surface, or line pre-existing fractures in the rock. Rock fulgurites are typically found on mountain tops, most famously on Oregon's Mount Thielsen, known as the Cascade's Lightning Rod due to evidence of strikes found there.

Rock type: Non-foliated, contact metamorphic
Texture: Glassy
Structure: Sand fulgurites: branching tubes of glassy sand in loose sand. Rock fulgurites: glassy crusts on solid rock in veins or fractures.
Colour: Grey-green to black
Composition: The silica mineral lechtalierite
Protolith: Sand fulgurites form in loose sand; rock fulgurites can form on any rock
Notable occurrences: Sand fulgurites: Sahara Desert, Africa; Namib Desert; Botswana; Lake Michigan, Atlantic coast of North America; Utah deserts. Rock fulgurites: Isle of Arran, Scotland; Mt Blanc (Alps), Pyrenees, France; Mount Ararat, Turkey; Toluca, Mexico; Sierra Nevada, CA; Wasatch Range, UT; Mount Thielsen (Cascade Range), OR; South Amboy, NJ

Identification: Sand fulgurites are branching knobbly tubes of glassy sand.

SPACE ROCKS

Meteorites are chunks of rock from space, mostly asteroids, that crash into the Earth. Because large meteorites strike the ground with such force that they are instantly vaporized, most meteorites are small – the largest ever, from Grootfontein in Namibia, is no larger than a double bed. But you can also find entirely new rocks and minerals forged by the impact called impactites, which include tektites and suevite.

Stony meteorite

Perhaps reflecting Earth's stony mantle and iron core, meteorites are divided into two main kinds, stony and iron, plus a few made from both stone and iron. Nine out of ten meteorites falling to Earth are stony meteorites, made largely of silicates and containing many minerals familiar from Earth rocks such as olivine and pyroxene. Stony meteorites are of two kinds: chondrites and achondrites. Chondrites get their name because most contain little globules called chondrules, typically 1mm/0.04in across. Most achondrites, on the other hand, contain no chondrules. Like grains in sedimentary rock, chondrules in chondrites are set in a matrix of finer material. The theory is that chondrules are droplets of olivine and pyroxene that condensed and crystallized in space while the Solar System was forming, then clustered together to create asteroids. The chondrites are the most jumbled in composition of all meteorites, perhaps little changed since the very earliest days of the Solar System. Minerals in achondrites are less jumbled, reflecting how they have become differentiated over time as asteroids and planets developed crusts and mantles. Most achondrites come from asteroids, but 28 have come from Mars and 20 from the Moon.

Identification: A single unusual-looking knobbly stone quite unlike any others in the area could just be a meteorite. Look for a dark outside showing signs of melting, and a light-coloured inside. A very heavy feel is also a good clue. If it is full of holes, it is more likely to be volcanic. Light-grey chondrule spots may be visible in a chondrite.

Rock type: Meteoritic
Texture: Fine-matrix, possibly with tiny spheres called chondrules, up to pea-size
Structure: No obvious internal structure
Colour: Light to dark grey, black
Composition: Similar to peridotites or gabbros, mostly made of olivine, pyroxene and nickel-iron. Chondrules are olivine, pyroxene, bronzite, diopside or, more rarely, chromite, magnetite, graphite or spinel. The matrix is the same material.
Origin: Asteroids and comets
Notable occurrences: Bjurbole, Finland (1899); Jillin, China (fell 1976); Hoba Farm, Grootfontein, Namibia (prehistoric times); Norton County, Kansas (1984); Long Island, New York (1948); Paragould, Arkansas (1930)

Iron and stony-iron meteorites

Iron meteorites are unlike any Earth rock. Thought to come from the cores of asteroids, they are almost pure metal – basically an iron-nickel alloy in the form of the rare minerals kamacite and taenite. Only 1 in 10 meteorites is iron, but they are much easier to spot being large, dark, heavy and odd-shaped. They also survive for a long time in soil. So, despite their rarity, they are found more often than stony meteorites, and all the biggest specimens are iron. They may be divided into three groups according to the structures that appear on their surface when etched with nitric acid: octahedrites, hexahedrites and ataxites. Octahedrites are marked by criss-cross ribbons called Widmanstätten figures and hexadrites by parallel 'Neumann' lines. Ataxites have no clear marks. Each group is characterized by a different mix of kamacite and taenite. The rare stony-iron meteorites are made of iron and silicate minerals. They are thought to come from the core-mantle boundary of large asteroids.

Identification: Knobbly, heavy, solid and unbreakably metallic – and magnetic – iron meteorites are easy to identify. Prehistoric people used them as a source of iron, called sky iron.

Rock type: Meteoritic
Texture: Fine-matrix, possibly with tiny spheres called chondrules, up to pea-size
Structure: See text
Colour: Brown, grey, black
Composition: Iron and nickel as two main minerals: kamacite (nickel-poor) and taenite (nickel-rich). Hexahedrites are mostly kamacite and ataxites mostly taenite. Octahedrites are both kamacite and taenite.
Origin: Asteroids and comets
Notable occurrences: Sikhote-Alin, Russia (1947); Odessa, Ukraine; Nantan (Guangxi Province), China (fell 1516, found 1958); Antarctica; Campo del Cielo, Argentina; Allende, Mexico; Canyon Diablo, Arizona

Tektite

Tektites are remarkable dark, glassy lumps, first observed by Charles Darwin in Tasmania. They are clearly made from molten glass, and Darwin thought they were volcanic bombs. Their origins sparked heated debate, but geologists now agree that most, if not all, tektites are solidified splashes of rock melted by the impact of a meteorite. They are usually found only in particular regions, called strewn-fields. Some strewn-fields are linked to known impact craters, like the Bosumtwi crater in Africa's Ivory Coast, and the Ries crater in Germany. With others, the crater has yet to be found. By far the largest field is in Indochina, stretching from Malaysia to Tasmania, where millions of tektites have been found. Particular kinds of tektite are named after a strewn-field, such as australites from Australia and the beautiful green moldavites from the Moldau region in the Czech Republic, used as jewellery in prehistoric times. There are four main kinds of tektite: microtektites are grain-sized balls found only in marine sediments; Muong-Nuong tektites are pea- to truck-sized chunks; australites have been shaped while molten into saucer shapes; and splash-form tektites are dark globs of black or green glass.

Splash-form tektite: These tektites are dark globs of black or green glass. They can be round-, teardrop-, disc-, dumbell- and even rod-shaped. Their surface is always marked by pits and furrows created by corrosion.

Rock type: Impactite
Texture: Glassy
Structure: See text
Colour: Black to green, occasionally yellowish
Composition: Similar to granites, with 70% silica
Origin: Meteorite impact melting sandstone and shale
Notable occurrences: Ries-Nördlingen, Germany; Moldau, Czech R (linked to the Ries impact); Irghiz, Russian Fed; Bosumtwi, Ivory Coast; Muong-Nong, Laos; Cambodia; Malaysia; Philippines; Mt Darwin (Tasmania), Kalgoorlie (WA), Victoria, Australia; Albion I (Corozal), Belize; Beloc, Haiti; Arroyo el Mimbral, Mexico; Bedias region (Fayette County), Texas; Martha's Vineyard, Massachusetts; Georgia

Meteorite impact sites
The Earth is struck by debris from space with remarkable frequency. Although smaller fragments burn up on their way through the atmosphere, at least three million meteorites big enough to create a crater at least 1km/0.65 miles across have crashed into the Earth during its existence. The effect of such impacts is clearly visible on the pitted surface of the Moon, but the Earth is so geologically active that the signs of many impacts have long since been wiped away here. Until recently, scientists believed that Earth was hit by meteorites only early in its history. Now they know it is being hit all the time, and the remnants of even quite ancient craters, or 'astroblemes', are there if you know what to look for – including key minerals such as stishovite. The first firm identification was Meteor Crater in Arizona, initially suggested as an impact crater by Daniel Barringer in 1902, then confirmed by Eugene Shoemaker, Edward Chao and Daniel Milton in 1960. Now, more than 160 major astroblemes have been discovered, including the Sudbury impact in Ontario and the Ries-Nördlingen site in Bavaria in Germany.

Suevite

Besides sending out splashes of molten rock, the tremendous impact of a meteorite can literally pulverize the rock where it hits creating what is called an impact breccia. One of these breccias, first identified at the Ries crater in Germany, is suevite. Suevite is a jumbled mix of glass melt bombs and fragments of crushed rock. Unlike other impact breccias, it usually forms only when there is plenty of water around. This water is usually in the ground, but the surprisingly large amounts of suevite at the famous Sudbury impact crater in Canada may be explained if the impact was not an asteroid, as most impacts are, but a comet. Comets bring water with them. Thin layers of suevite have proved to be a key piece of evidence supporting the theory that the dinosaurs were wiped out by the effects of a huge meteorite impact 65 million years ago, at Chicxulub in Mexico.

Rock type: Impact breccia
Texture: Breccia, with powdery matrix containing various-sized angular rock fragments and rounded glassy melt bombs
Colour: Buff, light grey with dark stones
Origin: Meteorite impact pulverizing country rock
Notable occurrences: Ries-Nördlingen, Germany; Popigai, Kara, Siberia; Lonar, India; Vredefort, South Africa; Bosumtwi, Ivory Coast; Woodleigh, Gosses Bluff, Australia; Chicxulub, Mexico; Haughton (Devon I), Nunavut Territory of Canada; Manicouagan, Quebec; Sudbury, Ontario; Manson (Des Moines), Iowa; Chesapeake Bay, Maryland and Virginia; Meteor Crater (Painted Desert), Arizona

Identification: With its mix of rock powder and glass bombs, suevite looks very much like volcanic breccia but is closely linked to impact sites.

GLOSSARY

accessory any mineral not essential to the rock's character.

accreted terrane belt of rock welded to the edge of a continent by subduction.

accretion various meanings, including gradual growth of a body such as a grain by addition of new material to the surface.

acidic rock rock rich in silica.

aggregate mass of rock or mineral particles.

allochromatic owing its colour to impurities.

alloy a manmade combination of two metals.

alluvial deposited by rivers.

alpine cleft mineral-rich fissure typical of the Alps.

anion negatively charged ion.

aphanitic of grains too fine to see with the naked eye.

arenite sedimentary rock with sand-size grains.

assemblage zone strata dated by a group of fossils

association minerals that commonly occur together.

asthenosphere hot, partially molten layer of the Earth directly below the lithosphere.

astrobleme eroded remains of a meteorite impact crater.

banded iron formation narrow iron-rich layers of rock.

batholith large mass of usually granite plutons at least 100 sq km (39 sq miles) in area.

bedding plane the boundary between layers of sedimentary

rock formed at different times.

bedrock solid rock that lies beneath loose deposits of soil and other matter.

Benioff-Wadati zone sloping zone of earthquake centres in a subduction zone.

biogenic formed by living things.

biostratigraphy dating layers of rocks with fossils.

carat unit of weight for gems, equivalent to 200 milligrams.

cation positively charged ion.

cementation the stage in lithification when cement glues the sediment particles together.

chalcophile of elements such as copper and zinc that have an affinity for sulphur.

clast fragment of broken rock.

clastic sediment sedimentary rock made mainly of broken rock fragments.

cleavage the way a mineral breaks along certain planes.

compaction stage in lithification when water and air is squeezed out of buried sediments by the weight of overlying deposits.

concretion distinct nodules of materials in sedimentary rocks.

contact metamorphism when rocks are metamorphosed by contact with hot magma.

contact twin twinned crystals in which each crystal is distinct.

continental drift the slow movement of the continents.

country rock the rock that surrounds a mineral deposit or igneous intrusion.

core the centre of the Earth.

craton ancient part of a continent unaltered for at least one billion years.

crust the top layer of the earth, attached to the upper mantle.

cryptocrystalline with crystals so small that they cannot be seen even under an ordinary microscope.

crystal form the way in which the different faces of the crystals are arranged.

crystal habit the typical shape in

which a crystal or cluster of crystals grows.

crystal system one of the different groups into which crystals can be placed according to how they are symmetrical.

cuesta low ridge with one steep side and one gently sloping.

D" or D double prime the transition zone between the Earth's mantle and core.

Dana number number assigned to each mineral according to the classification system devised by James Dwight Dana.

detrital made of rock fragments.

diagenesis all the processes that affect sediments after deposition including compaction and lithification.

diaphaneity the degree to which a mineral is transparent.

disseminated deposit mineral deposit created by infilling pores and cracks in igneous rock.

drift geology the geology of loose surface deposits.

dyke sheetlike igneous intrusion, either near vertical, or cutting across existing structures.

Ediacaran the geological period lasting from about 600 to 542 million years ago. This was added to the system in 2004.

effusive volcano volcano that erupts easily flowing lava.

element the simplest most basic substances, such as gold, each with its own unique atom.

Era vast portion of geological time lasting hundreds of millions of years.

evaporite a natural salt or mineral left behind after the water it is in has dried up.

exposure where a rock outcrop is exposed at the surface.

extrusive igneous rock type of rock that forms when volcanic lava cools and solidifies.

facies assemblage of mineral, rock or fossil features reflecting the conditions they formed in.

fault a long fracture in rock along which rock masses move.

feldspathic rock containing feldspar.

felsic rock rich in feldspar and silica, typically light in colour.

fissure volcano volcano which erupts through a long crack.

flood basalt plateau formed from huge eruption of basalt lava from fissure.

foid abbreviation of feldspathoid.

foliation flat layers of minerals in metamorphic rock formed as minerals recrystallize under pressure.

forearc basin region on the trench side of an island arc in a subduction zone.

fossil correlation cross-checking of fossils between separate rock outcrops, used in rock dating.

fractionation the way in which the composition of a magma changes as crystals separate out when it melts and refreezes.

fracture the way in which a mineral breaks when it does not break along planes of cleavage.

geode hollow globe of minerals that can develop in limestone or lava, often lined with quartz.

geological column diagram showing the successive layers of strata that have formed over geological time, with the oldest at the bottom, youngest at top.

glacial a period in an Ice Age when ice sheets spread, or anything related to glaciers.

graded bedding layers of rocks which show a decrease of grain size, from coarse at the bottom to fine at the top.

granular texture of rock with visible, similar-sized grains.

groundwater water existing below ground. See also *phreatic water*.

hydrothermal related to water heated by magma.

hydrothermal deposit mineral deposit formed from mineral-rich hydrothermal fluids.

hypabyssal igneous rocks in small intrusions such as dykes.

idiochromatic mineral getting its

colour from its main ingredients.

igneous rock rock that has solidified from molten magma.

impact crater crater formed by the impact of a meteorite.

index fossil key fossil used for correlating strata.

index mineral mineral that typifies a metamorphic facies.

intraclast any sedimentary rock fragment that originated within the area of the rock's formation.

intrusion emplacement of magma into existing rock.

ion atom given an electrical charge by gaining or losing electrons.

island arc curved chain of volcanic islands in a subduction zone, eg the Aleutians.

isostasy the natural buoyancy of the Earth's crust, making it rise and sink as its weight changes.

isotope variety of atom of an element that has a different number of neutrons (uncharged particles in its nucleus).

joint crack in rock created without any appreciable movement on either side, often at right angles to the bedding.

karst typical limestone scenery characterized by caverns, gorges and potholes.

kimberlite igneous rocks rich in volatiles, normally forming pipes.

Large Igneous Province Vast outflow of lava, typically on the ocean bed.

laterite weathered, soil-like material in the tropics rich in iron and aluminium oxides.

lava erupted magma.

leucocratic of light-coloured igneous rocks.

lithification change of loose sediments into solid rock.

lithosphere the rigid outer shell of the Earth containing the crust and upper mantle, broken into tectonic plates.

lode cluster of disseminated deposits.

lutite sedimentary rock made mainly of clay-sized grains.

mafic of rocks rich in magnesium and ferric (iron) compounds, equivalent to basic.

magma molten rock.

magma chamber underground

reservoir of magma beneath a volcano.

mantle the zone of the Earth's interior between the crust and the core, made of hot, partially molten rock.

mantle convection circulation of material in the magma driven by the Earth's interior heat.

mantle plume long-lasting column of rising magma in the mantle.

massive of rocks with even texture, or a body of a mineral without any distinct crystals.

melt a mass of liquid rock.

metamorphism the process in which minerals in rocks are transformed by heat and pressure to create a new rock.

metasomatism metamorphic process in which minerals are transformed by hot solutions penetrating rock.

meteoric water water that comes from the air, typically in reference to groundwater.

meteorite a chunk of rock from space that reaches the Earth's surface.

Mohs scale the scale of relative hardness of minerals devised by Friedrich Mohs.

MVT Mississippi Valley Type mineral deposits, formed as limestone is altered by hot solutions.

native element element that occurs naturally uncombined with any other element.

nodule rounded concretion.

nuée ardente fast-moving clod of scorching ash and gases created by a volcanic eruption.

ocean spreading the process in which oceans widen as new rock is brought up at the mid-ocean ridge.

oolith small, usually calcareous accretions in a rock.

oolitic made mostly of ooliths.

ophiolite grouping of mafic and ultramafic igneous rocks, including pillow lavas, that were once part of the sea floor.

ore natural material from which useful metals can be extracted.

orogeny mountain-building.

outcrop area of rocks occurring at the surface.

oxidation zone upper layer of mineral deposit where minerals are altered by oxygen and acids in water.

pegmatite very coarse-grained igneous rock, usually found in veins and pockets around large plutons and rich in rare minerals.

penetration twin mineral crystal twin in which the crystals have grown into each other.

Period major portion of geological time lasting tens of millions of years.

permeable of rock that lets fluid or gas to seep through it easily.

phaneritic of grains visible with the naked eye.

phenocryst relatively large crystal in igneous rock.

phreatic water groundwater in the saturation zone of rock below the water table.

pillow lava lava formed on the sea bed consisting of pillow-shaped blobs.

pisolith pea-sized, usually calcareous accretions.

pisolitic made mostly of pisoliths.

placer deposit of valuable minerals washed into loose sediments such as river gravels.

pluton any large intrusion.

polymorphs different minerals created by different crystal structures of the same chemical compound.

porous full of voids (holes).

porphyritic of an igneous rock containing lots of phenocrysts.

primary mineral mineral that forms at the same time as the rock containing it.

protolith the original rock forming a metamorphic rock.

pseudomorph mineral that takes the outer form of another.

pyroclast fragment of solid magma plug ejected during a volcanic eruption.

radiometric dating the dating of rocks from radioactive isotopes within it.

regional metamorphism metamorphism over a wide area typical of fold mountain belts.

rift valley trough-shaped valley bounded by parallel faults.

rudite sedimentary rock

consisting mostly of at least gravel-sized grains.

schistosity banding of minerals in schists created by parallel growth of mineral crystals.

secondary mineral mineral formed by alteration of primary minerals.

sediment solid grains that have settled out of water.

seismology study of earthquakes.

serpentinization alteration of ultramafic rocks to serpentine by hydrothermal fluids.

siderophile of elements such as cobalt and nickel that have an affinity for iron.

silicic of igneous rock such as granite rich in silica, making it acidic.

sill sheetlike igneous intrusion, either near horizontal, or following existing structures.

skarn typically mineral deposit created by the alteration of limestones by metasomatism.

streak mark made by a mineral rubbed on unglazed porcelain.

strike direction of fold or fault.

subduction zone boundary between two tectonic plates where one plate descends into the mantle beneath the other.

tectonic plate one of the 20 or so giant slabs into which the Earth's rigid surface is split.

tenacity how a mineral deforms.

turbidity current swirling undersea current.

twin paired mineral crystals growing together.

vein thin deposit of minerals formed in cracks.

vesicle small cavity formed by gas bubbles in lava flow.

water table level below which rock is saturated by groundwater.

INDEX